'In this love letter to the Continent, Robert Winder takes a meandering and always entertaining journey down Western Europe's three great rivers. His affection for his subject shines through on every page . . . and his daring digressive style gives him latitude for quirky erudition and the chance to deliver unexpected delights.'

Stephen O'Shea, author of *The Alps*

'Insightful, elegant, and deeply researched to the very last drop, this book will change the way you see rivers – and the continent they have shaped – forever.'

Matt Gaw, author of *In All Weathers*

'*Three Rivers* is wonderful excursion into history, travel and stories about some of the most fascinating rivers of Europe – do yourself a favour and take a holiday with this book.'

Robert Twigger, author of *Red Nile*

THREE
RIVERS

ROBERT WINDER

The extraordinary waterways that made Europe

Elliott&Thompson

First published 2025 by
Elliott and Thompson Limited
2 John Street
London WC1N 2ES
www.eandtbooks.com

Represented by:
Authorised Rep Compliance Ltd.
Ground Floor, 71 Lower Baggot Street
Dublin, D02 P593
Ireland
www.arccompliance.com

ISBN (hardback): 978-1-78396-896-1
ISBN (trade paperback): 978-1-78396-925-8

Picture credits
Page 115: Imagebroker.com / Shutterstock.
All other photography courtesy of the author.

Maps: JP Map Graphics Ltd

9 8 7 6 5 4 3 2 1

A catalogue record for this book is available from the British Library.

Typesetting by Marie Doherty
Printed by CPI Group (UK) Ltd, Croydon, CR0 4YY

To Hermione, Luke and Kit

CONTENTS

MAPS OF THE RIVERS

The three rivers, from source to sea. The star marks
the place where their journeys begin.

The Rhine

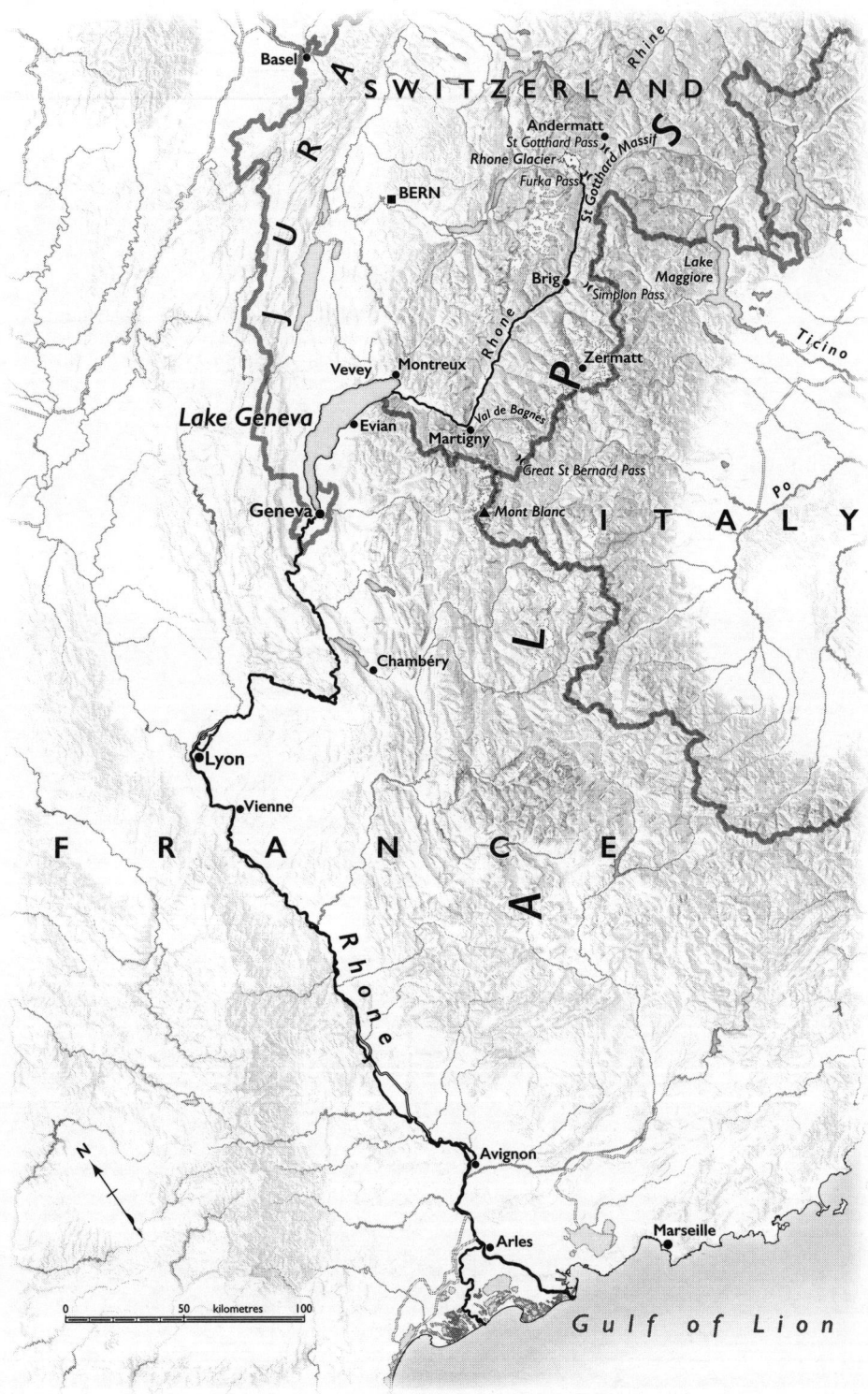

The Rhone

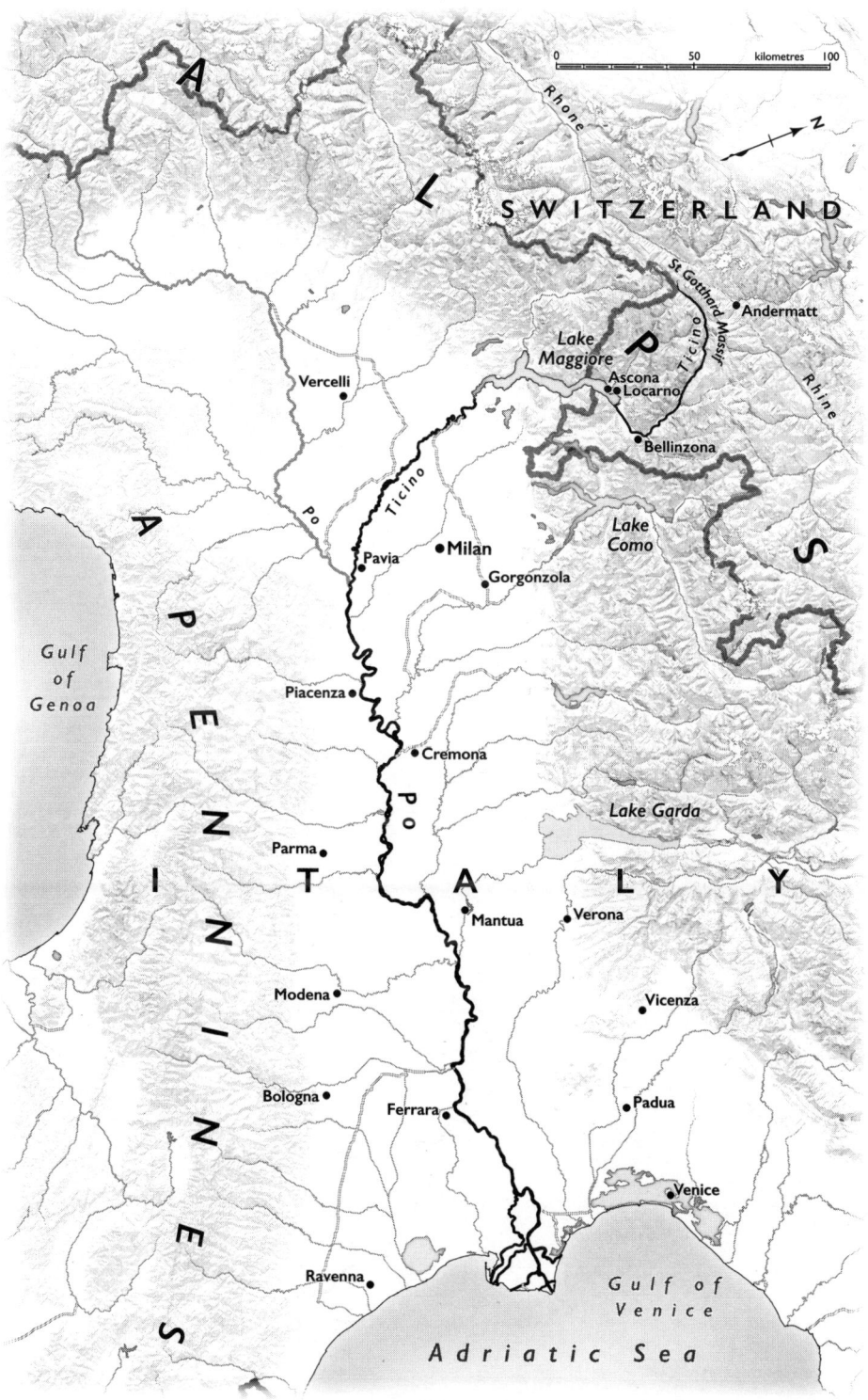

The Ticino and Po

1

THE CROSSROADS

If any one place could claim to stand at the heart of Europe, then the picturesque Swiss city of Basel might make a good candidate.* On the surface it seems a less than epic choice – prosperous and secretive, rarely awash with fluttering banners and bristling legions, like Rome or Paris, and not often stirred by the memory of marauding fleets, like London or Madrid. It makes most people think of banks, watches and chocolate rather than cultural star power.

But there is more to it than that.

Not least is the simple matter of its geographical location. If we drew a line running from London to Rome on one axis (via Paris) and another between Berlin and Madrid on the other, Basel would be close to the X in the middle. Wedged in the north-west corner of Switzerland, on the triple border of three different lands, peoples and traditions, it contains hints and pieces of all of them.

Visitors arriving at the airport have a choice of three exits (to France, Germany or Switzerland), and those heading for the former must pass along a narrow strip of fenced-in Swiss soil, fighting the temptation to breathe in, so great are the civilisations on either side.

---

* For the sake of a game – obviously there is no such spot.

Then they reach a well-groomed city centre crammed with corporate HQs, art galleries and high-priced restaurants, all of which look in several directions at once.*

Basel's most important feature, however, is that it sits on Western Europe's longest and most incident-packed river: the Rhine. A cafe on an island, no more than a ten-minute tram ride from the cathedral, offers views west towards Paris, north to Strasbourg, east over the Black Forest and south to the Alpine peaks that are the fabled emblems of Switzerland. Three countries can be admired in the course of a single cup of coffee, and a missile-shaped monument, the *Dreiländereck*, marks the spot where they meet and overlap. The fact that each of the three landscapes looks very much the same as the others is suggestive: it is hard to imagine a better reminder of the way in which the differences between Europe's nations, however clear and clung to, disguise a profound similarity.

The stream rushing past is far from ordinary. This is where the Rhine takes a hard right turn as it gathers strength for its run north. Between the steep walls it barges, shoulders hunched and head down, refusing to let anything stand in its way – a colossus, broad as the Thames at Tower Bridge but far less sluggish, a writhing monster heaving past the glass walls of Basel's towering office blocks, impatient for the castles and vineyards up ahead. But unlike the Thames, by that stage flavoured with mud, salt and seagulls, the Rhine still has 400 miles to go before it reaches the sea. And this fast-flowing water on the hinge of three nations, a major port in the heart of Europe's landmass, trembles too with hints of the three others on its mazy

---

* The railway station is in Switzerland, of course; a model of stylish efficiency. But platforms 30–36, offering departures to Dijon and Paris, are in France.

route – Austria, Liechtenstein and the Netherlands. That is not all of Europe, but it is a meaningful slice.

There are ancient echoes too: the Celts and the Romans built settlements here, popes held councils in these halls, and the huddle of medieval buildings up on the hill includes the university rooms where Desiderius Erasmus produced his Greek and Latin Bible, breathing new oxygen onto the candle that would illuminate the humanist revolution.

The thrumming power of this river delivers a neat reminder of a basic geographical fact: Basel is where it is *because* of the Rhine. There is no chicken-and-egg mystery about cities: people settled near water. The river came first. Nor is it a coincidence that Basel would become one of the cradles of European printing – a museum on the subject occupies a riverside mill to this day. That made it one of medieval Europe's busiest centres of radical thinking. It wasn't the river that translated the Bible or supported the scholars who gathered in the hilltop university (one of Europe's earliest) to dilate on the ideas and discoveries that drove the Reformation. But it *was* the river that watered the city that supported the libraries that gave Erasmus his intellectual platform, and it was the river that enabled the word to spread. The medieval *quartier* over which Erasmus presided was the crossroads of Europe, the intersection of all its pathways.

That is geography at its simplest: for centuries this city was the only bridge over the Rhine between the mountains and the sea, and it remains unusual in having a footprint on both sides of the river. Most of the Rhine's cities stand well away from its unreliable main channel, but thanks to its elevated position, Basel bestrides it with confidence.

Arguably, all of that is enough to make it a promising candidate for being the heart of Europe. But a few years ago, I started having second thoughts about it when I found myself on the Saint-Gotthard

Pass near Andermatt, 100 miles south-east of Basel. I was no longer on the expansive plain on the northern fringe of the Alps but on the high plateau whose winding roads – useless in winter – weave over the mountains into Italy.

To be exact, I was not *on* the pass so much as beneath it, in a railway tunnel, a mile below the mountain, staring at a map. I didn't have a view. But Andermatt, I now saw, lay on an even more suggestive junction than Basel. It was situated on the east–west road from Geneva to Salzburg and the north–south route from Zurich to Milan. It was the crossing point for traffic from Germany to Italy, and also the path from east to west. Maybe *this* was the crossroads, the junction that linked Europe with itself.

I frowned. According to the usual rules of human settlement, a pivot of this importance should have been busy with cathedral spires and castles. And what a place it might have been, a hybrid of the splendid cities (Lyon, Zurich, Innsbruck, Milan) on its spokes. But the terrain in these parts was too craggy and inhospitable for anything of that sort. Indeed, it might have been its relative isolation that allowed the importance of this snowbound spot to go so relatively unremarked. For centuries the Alps had seemed an obstacle or barrier to the easy flow of European life. Might it actually be the opposite, a critical point of intersection and exchange?

Then I noticed something else.

I had long known, after spending a student summer as a waiter in the Alps, that the source of the Rhone, which slices across Switzerland before tumbling down to the south of France, was only a stone's throw from the source of the Rhine, on the other side of the same mountain. But I had not noticed that three *more* rivers rose in much the same spot. Now, staring at a map on a subterranean train, the coincidence was hard to miss.

The Rhine at Basel: the watery hinge between three countries.

The first two, the Reuss and the Aare, are relatively short rivers – mere tributaries. But the third, which springs to life only a few miles south of my tunnel, is much more important: the Ticino.

It is not as well known as the Rhine or the Rhone, but in one major respect is every bit as evocative, because instead of flowing north to the Netherlands or west to France, this third body of water hurtles down into Lake Maggiore, then on past Milan to the hot blue haze of Italy.

That makes it important for two reasons. First, it was the route south, linking northern Europe to the civilisation of Rome. But also, after Milan, the Ticino makes its way to Pavia where it runs into the River Po. Linking arms, the two of them water the plain of Lombardy, irrigating the land all the way to Venice and the Adriatic.

The Po is Italy's longest river and is officially deemed to rise in the Cottian Alps west of Turin. But that is down to the geographical convention that measures by volume rather than by length. If such things were decided by distance, then the source of the Ticino could easily be thought of as the source of the Po, since its source is actually a few miles further from the confluence. It is not a competition – but it is the Alpine water flowing down the Ticino that transforms the Po into a mighty stream.

Which set me thinking.

Three major rivers, flowing in three different directions, carving out the three valleys that gave rise to three great civilisations – French, German and Italian – had to be worth a closer look. They were much more than mere scenery. Nor were they simply dumb witnesses to the long pageant of European history. They had been instrumental in building and decorating the stage on which it played itself out. And the coincidence of their shared origin near Andermatt depended on a geographical fluke as arbitrary as which side of a Swiss mountain the rain happened to fall.

Rivers flow downhill, and geography moves in unmysteri-
ous ways, so on reflection it is no surprise that all three of these
watercourses – Rhone, Rhine and Ticino/Po – should make their
appearance in the same patch of Swiss mist. But the scale of what they
have gone on to accomplish is remarkable. Contrary to the idea
that the Alpine passes were points of meeting and exchange, for the
first few thousand years of human habitation the mountains really
acted as walls, standing between distinct valleys in which people
developed markedly different ways of life. And since those French,
German and Italian strands were the three most instrumental voices
in European life, the broad civilisation of the whole continent (its
western part, at least) could in one sense be said to begin right here.
Spain and Britain might object, but it is not *very* controversial to
see the French, German and Italian parts of Europe as its primary
colours – or at least the loom on which those formative threads were
woven. And all of them begin life in this one Alpine massif.

The water trickling down the slopes above my head gave this
geological notion dramatic weight. To my right, the water ran west
to Lake Geneva, where it would pause to gather its breath before
spilling onwards into France. The water behind me, meanwhile,
was being ushered away to Germany, where it would become the
blood flow of *that* immense culture. Ahead of me, it was destined
to become Italian.

Given a clear day (and since this is a daydream we can assume
perfect weather) it would be easy to imagine a fantasy game of
Pooh sticks up here, dropping twigs into the three streams and
following them to see which arrived first in the North Sea . . . the
Mediterranean . . . or the Adriatic.

This mighty Alpine chain, I suddenly saw (from my tunnel) was
not just a photogenic playground for snowboarders and paragliders,

but a topographical sorting hat. And the scale of what its waters had carved out could hardly be overstated. The three great valleys were nothing less than the mould in which Europe had been made – geography's version of Adam Smith's economic 'invisible hand'. It was the shape of these basins that governed the whole pattern of human settlement. Their steep walls made it hard for people to venture sideways, out of their worlds, while the rivers encouraged them to move up and down. The valleys created bonds between neighbours while accentuating their differences from the strangers over the hill. The rest is, as they say . . .

Since the first roads – and later railways – inevitably ran on the level terrain alongside the rivers, the entire arrangement of Europe's cities, towns and villages, its cathedrals, castles, universities and factories, all of it, grew in the passages created by Alpine water. The same went for agriculture, which took root on the fertile soil beside the water; for transport, trade and tourism; and of course for the countless intellectual, religious, political and artistic currents that had swirled through Europe, using these same channels, for thousands of years.

That is not even to mention the battles they have witnessed, the religions they have fostered, or the industries they have empowered.

I was not the first person to notice the fact that the three rivers shared a single origin: there were already footprints on this mountain. The Alps had many times been referred to as Europe's water tank – or the tap on its roof. Swiss writers certainly knew that the Saint-Gotthard Massif was a historic meeting of the ways. In 1897 Carl Spitteler (a Lucerne poet who won the Nobel Prize for Literature in 1919) described it as Europe's 'central' hub, thanks to its 'whole range of spiritual and physical characteristics'. And many of his successors sang the same song. 'The St Gotthard,' wrote Emil Egli in *Swiss Life and Landscape* (1949), 'is the crossroads of Europe.' But though I was not the first, the growing modern preoccupation with all issues pertaining to climate, human folly and the health of the planet made it a particular geographical coincidence.

F. Scott Fitzgerald once called Switzerland 'a country in which very few things begin', and it must have sounded semi-plausible at the time. But the fact that the primary rivers of Europe were all fed by this one Alpine fountainhead made that seem a threadbare notion. All those leaps of thought and faith, the scientific breakthroughs, the electrifying currents of art, music and poetry, the whole tumultuous rainbow of Europe's many glories and crimes . . . all of it came from this one frozen cradle. Better to call it the place where *everything* began. Without the pattern laid down by these waters, Europe would simply not exist in the form we know it.

The only thing I could see was dark rushing stone – the cold innards of a mountain. But in my mind's eye I could see a fairy tale. From this angle, the Alpine mass above this tunnel, on the joint between German, French and Italian Switzerland, could have been the fabled rock struck by the wizard's staff, releasing in a flash the stream that alone could bring life to the parched landscapes below.

And then I was struck by a new notion. It was not enough merely to enjoy the fact that these three river systems of Western Europe began in the same glacial heights, or to smile at the arbitrary nature of the physical geography that determined their routes. I also wanted to explore the full broad magnitude of what they had accomplished on their roundabout journeys to the sea. It is impossible not to gasp at the scale and range of their footprints. These rivers had touched, fed and influenced every aspect of the varied civilisations through which they surged – the food, the art, the science, the religious colour, the industry, the wars, the revolutions – everything.

The intricate texture of Europe's first fruits (wine and cheese on the Rhone; meat and fish on the Rhine; grain and salt on the Ticino) had filled these valleys, along with everything that came after. These waters had been exploited for settlement, energy, transport, recreation and inspiration by millions of people, for thousands of years. Nations had risen and fallen, faiths had been born and died, wars had come and gone.

It would not be feasible to do more than sketch what the rivers have achieved – there is simply too much to take in. I would only be able to glance at the historical scenery, as if from the window of a racing train. But while tracing the courses of these continent-shaping waterways, I would find a particular story to tell – and these fleeting impressions might form themselves into a broadbrush collage of Europe's past and future.

It was time to take a closer look.

2

THREE VALLEYS

Before going any further, however, let us take a moment to orientate ourselves by travelling swiftly down the three rivers in question.

The first thing to appreciate is the plain geological fact of their shared birthplace. A million years ago (give or take a millennium or two) the Scandinavian ice sheet that coated Europe – north of a line running across England north of the Thames and then through the Netherlands, Germany and Poland – also blanketed the Alps. The mountains beneath its frozen dome had already been created by the collision of the African and Eurasian tectonic plates, which folded Earth's surface upwards. Running water carved valleys into the newly tilted surface, and the glacial heaps of ice, gathering boulders as they went, turned those riverbeds into canyons.

When the ice sheet melted (about 10,000 years ago) the streams running out of the glaciers grew stronger, and moved faster, but since some of the hills were composed of diamond-hard granite and gneiss, and some were made of softer limestone (the fossilised remains of aquatic life – this was once an ocean floor), the rivers were often diverted or deflected.

The Rhine headed east, as if its destiny were to meet the Danube and drive to the Black Sea, but a granite barrier turned it north to Germany. The Rhone set off in the opposite direction, but ran into the Alpine chain near Mont Blanc and was also pushed north. Had it found a way to stream on through, then Annecy or Chambéry might be the region's major city, with Geneva and Lyon just hidden-gem backwaters tucked away in the hills. The Ticino, meanwhile, curved along the southern side of the ridge, as if exploring its options before deciding which way to go before taking the plunge down to Lake Maggiore and Milan.

This is how accidents of geology determine the life of the world. Ice is transient, but it leaves indelible signs of its passing.

The furrows made by the rivers became first valleys, then worlds. John Ruskin, ruminating on 'the nature of Gothic' in *Modern Painters*, invited readers to imagine themselves as migrating birds, peering down as they flew north, watching the bright Mediterranean give way to the dark Germanic forest. We might experience a similar judder as we follow the Rhine from the Alps to the Netherlands, or drift down the Rhone to Provence, or float down the Ticino to meet the Po and the wide lush plain of northern Italy.

THE RHONE

Fifteen miles west of the Saint-Gotthard Massif, the Rhone blasts out of its glacier and charges west through the Canton of Valais towards France. It is a river in a hurry, enlarged by the cascades that topple into it on either side. This is classic glaciated landscape: a broad U-shaped valley walled in by sky-piercing heights – Bernese Oberland to the north (Eiger, Jungfrau, Mönch), Pennine Alps to the south (Monte Rosa, Matterhorn, Mont Blanc). In fact, the whole

12

Source of the Rhone: where the glacier becomes a river.

region was chiselled out by the original Rhone glacier, and the modern river hurtles between stony banks all the way to Lake Geneva.

A few miles short of that lake it is forced north by the ramparts of the Mont Blanc Massif and the Dents du Midi (products of a seismic brawl in prehistoric times) before squeezing up through the bottleneck at Bourg St Maurice. This is the narrow passage in which St Maurice, a fifth-century Roman soldier and early Christian, refused to engage (and slaughter) an insurgent Alpine army. The enraged Emperor Maximilian commanded that Maurice and his disobedient legion be executed, and in death the sensitive soldier became a hero, a martyr and a saint.

An abbey was founded on the spot where he fell, and in due course he gave his name to an Alpine spa (St Moritz), an Indian Ocean island (Mauritius) and an African nation (Mauretania). He became the patron saint of the Duchy of Savoy, and founding father to various chivalric houses – not to mention 600 religious foundations across Europe.

The pass he refused to defend still stands above the transport routes (river, road and railway) that feed through Bourg St Maurice like the wires in a junction box. And its defensive importance was still apparent 1,500 years later, in the Second World War, when Switzerland took pains to fortify the pass against German invasion. Rows of anti-tank blocks still jut out like blackened teeth in the smooth green pastures beside the torrent.

Having forced its way through this obstruction, the river relaxes into the sleek croissant of Lake Geneva, where it supports a lavish stretch of elite real estate.* The northern shore is lined with Swiss

---

* Not far to the north is the picture-postcard village of Gruyères, a caricature of Swissness, which gives its name to one of the world's most renowned cheeses. While

enclaves – Montreux, Vevey, Lausanne, Nyon – while the southern fringe (though lacking the south-facing view of mountains) is French – Evian, Thonon, Yvoire. It is a miniature Riviera in the summer and a handy base for winter sports too, with Chamonix, Megève, Villars and Verbier within easy range.

Both shores enjoy the same Rhone Valley microclimate and the French language. They also share the blessings of French cuisine – though even the Swiss do not do French food quite like the French.

As it approaches Geneva, pausing to generate the dramatic white plume of its *Jet-d'Eau*, the lake narrows, as if preparing to be a river again. The city, though French-speaking, straddles an international border (its airport has two exits) and is awash with global overtones – world-leading hotel chains, banks and multinational organisations, with a slew of chocolate and watch brands. But the water seems to sense that it is about to change nationality. Above the lake it is Alpine – fast, shallow and cold. As soon as it leaves it feels different. It is highly regulated: a series of dams extract power, stabilise the level and control the current. But it is at last a proper river, with nothing of the canal or drainage chute about it.

The hills on either side – Jura to the west, Alps to the east – force it to zigzag down to Chambéry before pushing up towards Lyon, yet still it feels like a river trying to make a fast getaway. It blasts through gorges, twisting and wriggling in search of the cleanest route. Only after Lyon, where it accepts the waters of the Saone and proceeds with renewed strength, does it seem to exhale. From here it runs due south, swift and true, guided by the walls of the Massif Central

---

the fields and pastures around the castle are not directly reliant on the Rhone, it is the river valley that makes the weather that feeds the grass for its famous cows.

and the Alpes-Maritimes, and with little to hinder it, on it gushes, through ancient Provençal cities such as Avignon and Arles, before bursting into the sea near Marseille.

THE RHINE

The Rhine is the longest of the three rivers, and in truth passes through so many worlds that it might better be seen not as one but as six separate streams. The first, the Alpine Rhine, begins in a lake only a few miles from the Rhone glacier. And while the Rhone flies west to France, the Rhine sets off in the opposite direction, towards Austria. And unlike the Rhone, which marches down an ice-scraped canyon, the Rhine bounds through a recognisably riverine landscape, with intersecting bluffs marking the course of the slaloming water. Near the ski resort of Flims the river twists through the cliffs left behind when the mountain collapsed, leaving a traffic-free forest, with a pretty red train to give people a taste of the scenery.

Above the polished roads, swishing with luxury cars, wedding-cake monasteries stand aloof on their outcrops, while chalets festooned with red geraniums overlook pastures tinkling with cow-bells and scattered with barns, chapels, woodpiles and memorial crosses. Footpaths twist up to the snow. It is a cliché of Swissness, freshly groomed with well-trained Alpine charm. It is no surprise that *Heidi*, by Johanna Spyri, was born and is set only an hour down the river, at the far end of this delightful dale.

In truth this stream, the so-called Vorder Rhein, is only half the story, because it has a near-identical twin – the Hinter Rhein – which foams out of the mountains a few miles south, at the foot of another pass – the San Bernadino. Those familiar with the German Rhine, a sea-lane, might find it hard to equate that drumming highway with

The Swiss Rhine tumbles through the meadows where Heidi once roamed.

these babbling rivulets, dancing like deer through the wooded glades and green pastures. But they are all the same river, descendants of the same mountain source.

The Vorder Rhein, the Andermatt branch, is usually recognised as the official fountain on the grounds that it carries a greater volume. It is also the longer of the two branches. But not by much: it measures only 5 miles more than its twin (47 rather than 42), so perhaps we should say that the Rhine has two parents, becoming its true self only when they meet. This they do at Reichenau, near Chur, where a castle and a church stand on a wooded island in the middle of a confluence linked by three bridges. For such a significant junction, it is not easy to find; there are few vantage points from which it can even be glimpsed, apart from the iron railway crossing and a muddy footpath in the trees.

But thanks to the sculpting power of these headwaters, Reichenau has long been a meeting place for the roads through these hills. The monastic trail from Zurich to Milan followed the valley floor past the abbeys of St Gallen and Disentis, and Reichenau (named after an island in Lake Constance) had its own Benedictine abbey in 724 CE. At a time when Britain was grinding through its post-Roman Dark Ages, this remote landscape was a well-established route for travelling pilgrims.

That the river twists north at this point is thanks to the Calanda Massif, a 9,200-foot mountain that directs the stream left. The elbow is guarded by the venerable little city of Chur, and it says something about the historic importance of the location that this is Switzerland's oldest town. This was the way to Italy, and there was a settlement here in the Bronze Age. It was the capital of a Roman province and a bishopric in medieval times, with a tall-spired cathedral that dates back to the twelfth century.

From here the river runs through Vaduz, the capital of Liechtenstein – beneath the so-called Alte Rheinbrücke, a wooden bridge built in 1901 – before marking the western fringe of Austria, through Feldkirch, where James Joyce was once arrested,* to Lake Constance – or, in German, the Bodensee. There it swivels west, a complete volte-face, and though the chameleon nature of Switzerland means there is no linguistic jolt, the lake itself contains three national personalities – Swiss, German and Austrian.

By the time it leaves the lake, the Rhine gives the appearance of having enjoyed an invigorating rest cure in one of the area's exclusive wellness clinics. It slides under the bridge at Konstanz as a bigger, brasher animal known as the High Rhine. After streaming past the medieval facades and watchmaking factories of Schaffhausen, it plunges over the Rhine Falls, the most powerful cascade in Europe and a long-standing tourist draw. In the old days these falls attracted cultivated admirers such as Ruskin and Turner. Now they welcome 1.5 million visitors a year, most of them willing to buy a Swiss-priced ice cream, coffee, beer, room or ferry ride.

In even older times – the fourth century – this stretch of the Rhine was the edge of the Roman Empire, bristling with military posts and watchtowers. According to the Greek historian Ammianus Marcellinus, there were fifty of these fortifications between Lake Constance and Basel, and they were hazardous places. The

---

* He was thought to be a spy and arrested at the railway station while passing through during the First World War in 1915. Years later he claimed that the underlying concept of *Ulysses* – the whole silhouette of a wandering homesick traveller – crystallised at that moment. 'Over there, on those tracks,' he wrote, 'the fate of *Ulysses* was decided.'

small cadres of half a dozen men manning the towers knew that they were too weak to resist a concerted attack, so when 300,000 Teutons crossed the river in 378 CE, nothing stood in their way.

The Visigoths had arrived.

At Basel, as we have seen, the Rhine enters its third or 'Upper' phase – a misleading term arising from the fact that the earliest measuring devices recorded distances only from Lake Constance. In those days the river to the north was free-running and unreliable, and to this day it remains a treacherous flood plain, with no major conurbation until Mainz (the likes of Mulhouse, Colmar and Freiburg are all set prudently back). It is drenched in European history, though. It includes the cathedral spires of Worms, Speyer and Strasbourg, along with their vineyards and orchards, but also Enlightenment capitals at Heidelberg, Karlsruhe and Mannheim.

After Mainz we enter the Middle Rhine: the spiritual heartland of the Holy Roman Empire, home to the curving sweeps of water beloved of the cruise-ship industry – all ruined castles and Wagnerian myths. To call this stretch of the river operatic is to understate things. It has overflowed with the atmospherics of love and death for two thousand years. In between the pretty vineyards and gabled rooftops we glimpse bomb-blasted high streets and austere souvenirs. Here and there stand gleaming brass studs, planted in the cobbles of Bacharach, Heidelberg and other beauty spots to commemorate the Jewish families torn from their homes and trucked to their death in Hitler's Holocaust.

It is a river of inspiring strength but of dreadful memories too.

After Bonn and Cologne there is another mood swing. The Lower Rhine is a plough horse, and its cities smell of coal, diesel and petrochemicals. Leverkusen, Duisburg, Ludwigshafen: these are the groaning muscles of German manufacturing, and the Rhine is their lymphatic system. Up and down the barges thrum – and the freighters, and the tankers – plying their blunt-nosed trade on this grinding carriageway. Some have names like *Matterhorn* or *Rothorn*, stained reminders of the heights from which the river has come. But there is nothing in its oily current to suggest an Alpine spring. What began as a torrent has become a superhighway.

And it is the last stretch that belongs to Germany. From here on it is the Nether Rhine, the river delta otherwise known as Holland. At Arnhem, the 'bridge too far' that resisted Montgomery's airborne assault in 1944, it crosses a border, enters a new language, and takes on a new character. It also splits into three: the Waal, the Meuse and the Scheldt. Historically they were hard to control and often flooded. Only in recent times have they been tamed. One flows into the sea at Rotterdam, Europe's largest port, a last reminder that this whole plain is really the mouth of a river.

THE TICINO

We must not hold it against the Ticino that it is the least dramatic of our three headwaters. The Rhone and the Rhine have clear origins in a glacier or a lake. The Ticino makes a less theatrical entrance. It filters out of a rarely visited slope below the Nufenen Pass, which links the Rhone Valley to the southern flank of the ridge above the Val Bedretto in Italian-speaking Switzerland. But this is no crime: some of the world's greatest waterways have confused sources. The exact origin of the Amazon is contested – it wanders out of a watery maze in the Andes. The Nile used to announce itself boldly, bouncing out of Lake Victoria in the spray of the Ripon Falls – but in 1954 those falls were submerged by a new dam, which rather spoiled the effect. The Yangtze is the merger of several Tibetan rivers, and the Thames is a barely visible puddle in a field near Cirencester that actually dries up on sunny days.

That is how it is with the Ticino. When I went looking for the source in the summer of 2023, the clouds were low, visibility was about a yard, and the drizzle was thick. Apart from a motorcycle group that had pulled off the empty road for a rain-soaked cigarette there was no one about, and the source was hidden in the fog. There was a stone marker close by, but the Ticino is fed by all of the white streams coursing down the cliff, each of which (and none) has the right to feel pre-eminent.

This is still Switzerland (Canton Ticino) but it feels, to anyone coming over the pass, as though a major boundary has been crossed. The mood has changed – motorists who only five minutes ago were being warned to be careful (*Straßenschäden!* – uneven road surface) are now informed of a *Controllo della velocità!* (speed control). Once an important part of Rome's empire, this windswept upland was variously occupied by Ostrogoths, Lombards and Franks, and was an on-and-off fiefdom of the Duchy of Milan, too.

In 1478, 600 Swiss fighters stole over this height and laid siege to Bellinzona, the fortified city surrounded by orchards that guarded the road to Milan. The Duke of Milan sent a seemingly overwhelming force of some 10,000 troops to push them back, but they were lured into a gorge, ambushed from above and bombarded with heavy boulders. It was December, so the trap was sealed with deep snow. It was said that the crafty Swiss wore spiked shoes to improve their footing as they massacred 1,400 of their Milanese opponents. Exaggerated or not, it was a rout. The Swiss Confederation seized the all-important north–south road.

Ticino remained semi-independent until 1803, when, under pressure from Napoleon, it joined the Swiss Confederation and became

The Ticino trickles out of the mist in a fan of tiny streams.

part of a modern nation state. It was the last canton to join the union, and it made Switzerland a trilingual country with a footprint south of the Alps.

It makes for a beguiling fusion of Italian food, weather and cheerfulness with Swiss roads and plumbing. But the top of the Ticino valley is sparse. The villages – clumps of chalets – are tiny and feel empty. The pine-clad slopes leave little room for livestock; the air is too thin and the soil too sparse for effective farming. The steep walls make it dangerous, too – the village of Bedretto was half demolished by an avalanche in 1863, leaving twenty-nine dead. There is a bit of cross-country skiing up here, but no fashionable resorts. It is mostly a place people have wanted to leave.

As the river falls away to the south, the valley broadens into a fertile suntrap, ideal for fruit, corn, vines, tobacco and vegetables, while the water is tapped for hydroelectric power. Soon it reaches the scene of that medieval battlefield, just outside the region's capital, Bellinzona, a city that, like Sion and Chur, stands on a tall outcrop above the river, with three tall castles as souvenirs of its war-torn past. In 1878 it became Ticino's administrative centre, and it remains an important junction: this is where the roads to Italy part company, one following the river down to Lake Maggiore, the other racing away to Lugano. Both lead on to Milan.

The first thing the Ticino does after leaving Lake Maggiore is pour through the nature reserve* that sits alongside Milan's Malpensa Airport, watering the fields on its flank through an extensive lattice of canals and ditches. From there it runs to Pavia and the Roman-era

---

* The Parco Naturale Lombardo della Valle del Ticino, a sanctuary for herons and sparrowhawks. It also offers hiking, riding and kayaking on the river.

bridge on the confluence with the Po, where it forms the border between Lombardy and Piedmont.

The chain of cities that runs along the valley from there is by any standards magical: Piacenza, Cremona, Padua, Mantua, Ferrara. Each of these is a wonder in its own right; taken together they are a masterpiece.

None of them straddles the river. Like the Rhine, the Po was too big and moody. Set even further back are Parma, Modena, Bologna and Ravenna, while to the north stand Brescia, Bergamo, Verona and Vicenza. And out in its lagoon, incomparable on its stilts: Venice.

~

These, in brief, are the three valley civilisations of Europe. When people talk about 'the West', this is what they mean: the Graeco-Roman-Franco-German flow of Europe's past. And all of it was given life by the shape of the watersheds and the way the breeze happened to be blowing when the rain fell. We should not be surprised – history, it has been said, is only the product of geography over time. But the plod-plod of cause and effect over thousands of years has had enormous consequences. These glacial basins created three valleys, three lakes and, finally, three nations.

A fourth river hovers in the background of this story, larger than all the rest. And in one way the Danube does belong in this conversation, since it rises in the Black Forest of southern Germany, so close to the Rhine that some of its headwaters actually wander into Lake Constance. If we see the Rhine and the Danube as twin streams of the old Holy Roman Empire, they seem like cousins – children of some formidable dynastic marriage.

But although it rises in Germany, it streams eastwards into Slavic Europe, invigorating a different world: Slovakia, Hungary, Croatia, Serbia, Romania, Bulgaria, Moldova and Ukraine. It is the central thread of a different history – the drama of *Mitteleuropa.* That is why we shall not dwell on the Danube here.

Victor Hugo once wrote that 'each of the great Alpine rivers, on leaving the mountains, has the colour of the sea towards which it is flowing'. The cultures they support tell different stories and different jokes (in different languages), read different books, drive different cars, listen to different music, have different political systems, cherish a variety of beliefs and eat their own food. The Rhone is a land of olive oil, the Rhine a pasture full of butter, the Po alive with the aromas of Parmesan and prosciutto.

It is pleasing to think that rivers might possess in their youth aspects of their personality that emerge more fully later, that the child really can be father to the man, even in geography.* And the Rhine does carry the mud-salt tang of the North Sea, the Rhone does have a Mediterranean gleam, and the Po is indeed an irrigation trench for Lombardy. In the same vein, the Rhine tends to strike us as a river of liquid power – military, industrial and cultural – while the Rhone makes us think of crumbling walls, sun-soaked vines and rolling fields of lavender. The Po puts us in mind of Renaissance cities sitting in a sea of rice, sausage, grain and cheese.

In these lights it is natural enough to see the Rhine as brown, the Po as green and the Rhone as blue. But assertions of this sort rarely survive close scrutiny. Generalisations, as they say, are *always* wrong, and rivers are mutable, responsive to any shift of terrain or weather.

---

* An old saying that sounds patriarchal to modern ears. And while rivers were often seen as masculine (Old Man River, Father Thames, Le Rhone, Der Rhein), this is not universal: La Seine, La Loire. And in Oxford the gender-fluid Thames is Isis.

On cloudless days the ale-brown Rhine can glow like a sapphire, in winter the azure Rhone can be an icy shade of grey, and at sunset the Po glows gold. In the woods, all rivers are a leafy shade of green, and on dark nights they are black. The Rhine and Rhone have vines, to be sure; but also power stations; when the Rhone flows through a major city such as Lyon it is as tea-brown as the Thames. All three have Roman remains, and the Po has enough industrial blight to give it the worst air quality in Europe.

They all have more than one face, in other words. They are different, yet the same. Only by following them out of the mountains shall we be able to measure their characters and weigh their influence.

3

WHAT IS IT ABOUT RIVERS?

Why do rivers so beguile us? Those who live beside them speak of the way they become familiar, like friends or relatives ('Old Man River'), with moods as changeable as our own – cold and ruffled one moment, overweight and lazy the next; by turns bad-tempered or enfeebled.

Certainly, as I made my way up to the head of the Rhone, on the zigzag mountain road to where the glacier begins in a splash of white water, I found myself mesmerised. It seemed simultaneously permanent and fragile, fast-flowing yet steady, constantly renewing itself but within a single fixed form.

Rivers answer our basic need for water, shelter and soil. Deprive us of a drink for a day or two, and we perish. And since the world holds only one gallon of fresh water for every hundred of seawater, it is extremely valuable. That is why civilisation emerged on riverbanks – on the Nile, the Euphrates, the Indus and the Yangtze. The first Europeans settled on the Neander (hence Neandertal*), a river off

---

* *-tal* being German for 'land'. The valley was named after a seventeenth-century religious thinker named Joachim Neander, but it was the ancient bone fragments

the Rhine near Düsseldorf, and the superb agricultural conditions – good for meat, fish, vegetables and cereals – provided them with a diet so nourishing they could hardly fail to thrive. There was great excitement, in 1991, when the 5,000-year-old skeleton of a Neolithic man emerged from a glacier in the Rhone Valley.* The fact that he was on the ridge between Austria and Italy, and had what looked like an arrow wound in his shoulder blade, suggests that even in those far-off days he might have been defending a border.

It took time for the valleys to become populated away from the riverbanks. They were sheltered, but wrapped in cold, wolf-haunted forests that made them less inviting. Once the trees were cleared, the rich soil and easy climate did the rest. They have been the centre of European life ever since.

Rivers answer a philosophical thirst, too, in that their ambiguous nature – solid, yet fluent – invites contrasting responses. As the Greek seer Heraclitus (who famously saw nature as existing in a state of never-ending flux) wrote, 'No man ever steps in the same river twice . . . it's not the same river, and he's not the same man.'

From a distance they look stable, yet close up they brim with speed. They bulldoze the land, carving out space for towns, roads and railways, but also reclaim it – they are natural earth movers. They are placid, yet moody; smug and silver, like Shakespeare's Trent, or skittish, all *tirra-lirra*, like the stream that sent Tennyson's Lady of Shallot 'skimming down to Camelot' through groves of willow and aspen. As snaking paths of water that chisel their way through contoured landscapes, they bind people into fraternal groups. But they

---

found in 1824 – later identified as being roughly 40,000 years old – that made it famous. The bones themselves suggested humans with unusually large jaws.

* He was given the name Ötzi; his remains, the oldest in Europe, are displayed in Bolzano, Italy.

also divide them as barriers. And while in the most obvious way they sustain life, they can take it too.

It is a thrill to swim in them, feeling their weight and strength – and with their rocks and cliffs they offer adventure. But they are also havens of respite and repose. Their currents carry memories of fabled battles and epic river crossings. Rivers are both the path ahead and the point of no return. On the surface they appear purposeful, one-way arrows darting from the heights. But they are also fidgety: surging, eddying, sometimes narrow, sometimes broad, fast, slow, but never at rest, forking like forest paths. That is why, in our metaphors, they resemble stories or journeys – they have small beginnings, complicated middles and a variety of endings.* They are symbolic of time. Their one-way flow, predictable as a pulse, suggests that there can be no turning round. Yet heading upstream to a river's origins is like turning the clock backwards. As Joseph Conrad wrote, steaming up the Congo in *Heart of Darkness*, 'Going up that river was like travelling back to the earliest beginnings of the world.'

Few have charted this duality so well as the evolutionary scientist Stephen Jay Gould. In *Time's Arrow, Time's Cycle* he beautifully describes the rhythm of Earth as being represented by two contrasting metaphors: an *arrow*, following a relentless path, never repeating itself, and as a *cycle* of death and rebirth. The three rivers streaming out of the Alps were a picturesque fusion of these two visions – unstoppable forward motion within a larger cosmology.

---

* The 'course' of history is one of many metaphors we take from the life of rivers. We think in streams of consciousness, sometimes we meander or change horses midstream. And though at such times we might resemble fish out of water, and opt just to go with the flow, we know (of course) that still waters run deep, and that one day it will all be water under the bridge, or at least screened after the watershed.

Rivers are an integrated part of the water cycle, which sucks moisture out of the oceans, condenses it into clouds and drips it down again on land. They are splashing engines of perpetual motion, never at rest. Even when supposedly subdued by the taming hand of civilisation, they retain their wild nature. They seem singular, with their own relentless will; yet the way they hint at reincarnation and renewal makes them privy to eternal mysteries too.

Along their length rivers carry the fragrance of faraway places – some pastoral, some urban. They face both *up* towards their origins and *down* to their futures, and while they sometimes seem purpose-built for human pleasure – ideal for 'messing about' on – they are also places of heavy lifting and hard labour. They can appear light-hearted, babbling and careless, but just as often are emblems of persistence and constancy. They cut paths through wild landscapes, yet can also loiter and digress.*

Sometimes they stand for forgetfulness or indifference (they just keep rolling along) but at other times they are agents of memory, trembling not just with echoes of old battles but with artistic movements, foods, wines and famous names. In their time they have carried fleets of wine, coal, wheat, steel and oil, while witnessing a roomy anthology of myths and legends. They can be sacred (the Ganges, the Jordan, the rivers of Babylon) but also – more frequently now – case studies in despoliation.

Rivers are our servants, our water carriers – yet also our masters. We are bound to them, and for thousands of years have lived beside the mills that used their water to grind the grain we baked into our daily bread. Their slow, creaking music was for centuries

---

* The Severn, for example, Britain's longest river, rises just 15 miles from the Welsh coast, but obstinately trundles east out of the Cambrian hills before reaching the sea, 220 miles later, near Bristol.

the soundtrack of European life, and the miller is a local hero in many languages.

It could be argued that the world's grandest cities are ocean-facing ports – Athens, Lisbon, London, Mumbai, New York, Shanghai, Sydney and Tokyo. But some of the prettiest – Florence, Paris, Rome, Budapest – are on rivers. By comparison, the cities that have no significant running water – Beijing, Delhi, Madrid, Mexico City – always feel somewhat parched.

For a time, it was deemed old-fashioned to see unglamorous physical geography as being the force behind historical events; the world preferred to see human culture as 'imaginary' or 'constructed'. But today it is increasingly recognised that it was not abstract ideas that led people in the Arctic to live (and think) differently from those in the tropics. Culture has always been influenced, to a large extent, by geographical features – millions of lives have been moulded by the prevailing weather, their proximity to oceans or by which side of a river they were born. And the growing evidence that human life has harmed the natural world is reminding us of the reverse: that human nature too has been shaped by the stones that were once its cradle and have always been its home.

And by the rivers, of course.

4

OVER THE TOP

The first and perhaps most consequential of the rivers' many creations were the mountain passes around the cradles in which they were born. Millions of years of relentless erosion from ice and water (trickle-down geology) raked out the grooves down which the waters could fall, and in time these grooves became the sheltered valleys in which life could flourish. The saddles between the ridges at the top became crucial crossing points.

The Furka Pass between the Rhone and the Rhine, the Nufenen between the Rhone and the Ticino, and the Saint-Gotthard between the Rhine and the Ticino: all were critically important gateways, not just for the occasional early traveller but for European civilisation as a whole. So, before we go any further, let us pause in the heights of the Saint-Gotthard Pass, to consider the land that produced the rivers that shaped the worlds below.

The St Gotthard who gave his name to this pass was a tenth-century bishop from Bavaria who spent many years in Salzburg, Austria, and also travelled in Italy. He founded an Episcopal school at Hildesheim, near Hanover, and when he was canonised in 1131, his relics were associated with miracle cures. The *hospiz* on the pass (built in 1237, according to most accounts by the Archbishop of Milan) is an exposed yet sturdy link to the old Europe of chalices and vespers. St Gotthard became the patron saint of travelling merchants and is venerated in Switzerland – especially by motorists. That looping road down the south side of the ridge is perhaps the most dramatic of all the Alpine itineraries. The backdrop for a thousand car-promotion videos, it is *Top Gear*'s idea of heaven.

Inevitably, so renowned a crossing has attracted a great deal of military attention. But even in this respect the memorial on the pass is unusual. On a mound beside the car park stands a sculpture of an aged figure, bent double, leading a mounted cavalry officer. Most passers-by will take it to be a tribute to St Gotthard himself, but in fact it is the Russian Field Marshal Alexander Suvorov.

A name no longer on many lips, at the end of the eighteenth century Suvorov was the pre-eminent officer of the tsar's army, and also (such being the ways of the aristocracy at that time) a generalissimo of the Holy Roman Empire, a prince of Italy, a six-times-wounded prince of Sardinia, Count of Rimnikskiy, holder of the Order of the Black Eagle, the Golden Lioness and twenty other titles, and a Habsburg field marshal too.

He fought sixty-three battles and was never defeated, but this sculpture is in memory of the successful campaign he fought through these mountains in 1799, when he drove Napoleonic France out of Italy.* Suvorov led his 20,000-strong Russian army up the Ticino, joined forces with Austrian brigades, and pushed the French off the Saint-Gotthard Pass (at the cost of 6,000 men). He continued to pursue them down to the Schöllenen Gorge, near Andermatt, where

---

* He did, however, win battles at Trebbia and Novi, capture fortresses at Brescia and Mantua, and liberate important cities, including Milan and Turin.

Russia's Marshal Suvorov: still patrolling the Saint-Gotthard Pass.

the River Reuss plummets down a chasm crossed by the so-called 'Devil's Bridge', where he won another fierce fight.*

Switzerland has regarded him as a national saviour ever since.

It means that the presiding characters in this rugged mountain pass share the distinction of not being Swiss: one is Bavarian, the other Russian. And it explains one of Andermatt's unusual features. A ski resort in a neutral nation whose greatest fear would appear to be the climatic threat to its snow record, it is also a garrison town. Its grandest building is not a palace hotel or a railway station, but the Kaserne Altkirch army barracks.

A mural in the officers' canteen depicts Suvorov himself, hero of the Saint-Gotthard, standing in the snow outside the *hospiz* in September 1799.

~

Today there are three tunnels through the Saint-Gotthard. The first was for steam trains. Work began in 1872, using hand-held picks and shovels, and it opened ten years later. The second, for a motorway, was completed in 1980. And the one through which my train swayed at the start of this account was the 35-mile Base Tunnel, opened in 2016 after scooping out enough rock (the engineers liked to boast) to make *five* pyramids of Giza.

All three of these tunnels through the Saint-Gotthard were, at the time of their launch, the longest in the world. And in the 1870s, when the labourers on the first tunnel, who had started on opposite sides of the ridge, finally met after hacking their way through the

---

* After the war, the bridge saw service as an artist's model. J. M. W. Turner arrived there with his watercolours a few years later, in 1803–4. But that bridge is no longer alone – a road and railway line now cross the same deep canyon.

mountain towards each other, the divergence between them was so minute (1 foot in width, 2 inches in height) that Switzerland gained a reputation for civil engineering it has never lost.

Their timing was fortunate: the designers were able to exploit Alfred Nobel's labour-saving invention, dynamite (first demonstrated in 1876). It was dangerous: several hundred workers died during the tunnel's construction, 200 in a single accident. But it was finally completed, and opened in 1882. The opening ceremony was a national event, carefully curated as a beacon of national progress. Posters featuring William Tell, folkloric father of the nation, smiled over the launch, and a choir from Altdorf, the hero's birthplace, sang patriotic songs.

As it happened, Tell was a relative newcomer to the Swiss pantheon. Just as King Arthur lay lost in English fog until the Victorian era rediscovered him, Tell had been forgotten by his compatriots until Schiller's 1804 play made him a hero once more.

That play was to some extent inspired by this landscape, too, since it was here that Schiller first learned about Tell through his friend Goethe, who came across the story on his Swiss excursions. Goethe went on to direct the play's first performance in Weimar – Switzerland's self-image owes a surprising amount to Romantic German tourists. In 1775 Goethe actually trekked all the way up to the Saint-Gotthard Pass purely to enjoy the view. It was, he wrote, a place of 'desperation, struggle and sweat . . . bare as the valley of death, scattered with bones – fog – lake'.

The panorama to the south was another matter: the cypress and honey warmth of Renaissance Italy stretched as far as he could see. He made a quick sketch – the *scheide Blick* (farewell glance) – which stamped into the public mind the glorious view from this ridge between two worlds.

It is debatable whether the central event of the Tell legend – the father shooting the apple from his son's head with a crossbow – ever happened. Similar incidents feature in other European tales (it resembles Robin Hood's arrow-splitting gift). But national myths are not documentaries.

However, the very mobility of this story only confirms the sense in which this swirl of peaks and passes, far from being marginal to the flow of European history, a rock in the stream, is actually one of the places where it first started to come together.

5

ICE FLOWS

In Victorian times a Swiss entrepreneur built a hotel on the hair-raising road that coils to and fro beneath the Rhone Glacier, giving guests a close-up view of the ice wall. He called it the Belvedere, and it became a must-see destination for the grander sort of tourist – popes, monarchs, Hollywood stars – and also for the burgeoning coach-tour generation.

It even crops up in the background of a celebrated car chase. This is where James Bond followed Goldfinger's yellow Rolls-Royce through the Alps while being shot at by a well-spoken English 'It' girl in a sports car. In a barely noticeable continuity slip, the hotel actually pops up twice in the same sequence as the cars zoom by, but the scenery is magnificent – well worth a second glance. And it is possible to see that in 1963, when this scene was filmed, the glacier behind the hotel was still in evidence. It brought a delicious cold sparkle to the gentian-blue Alpine sky.

Not any more.

Today the hotel is a sorry spectacle – idle and vestigial, moth-balled and derelict, far below the glacier's grey snout. The ice has retreated up the hill: the hotel now presides over a stony valley and a switchback in the road. It was one of the creative inspirations behind

Wes Anderson's *Grand Budapest Hotel* (and it features as the cover image for his book of reminiscences). But it is no longer a functioning resort: its *raison d'être* has melted.

Modern ice-tourists park their cars and motorbikes, or climb out of their coaches (ironically, the devices most obviously responsible for eroding the sight they have come to see), and walk a few hundred yards along a scree slope to reach the 'ice grotto', a theme park drilled into the glacier's tongue. A wooden boardwalk, somewhat like a First World War trench, takes them into the blue interior. Since the whole thing is moving, the tunnel has been carved anew every year since 1870.

In modern times it is not the glacier's forward flow that matters so much as its retreat. The base is licking back up the rocky slope like a snake's tongue, at a rate of some 98 feet a year, and this makes the grotto resemble a marquee – like the glacier itself, it is wrapped in canvas in the summer in a vain attempt to slow the rate at which it is melting.

Beside the car park is a gift shop, a cafe, a kiosk that charges entry to the almost empty bowl, and a warning (not needed when the hotel was built) that drones are not permitted. A stone pipe splashes mountain water across the tarmac, some of which will tumble into the Rhone. The river itself is visible from the viewing platform, running due west towards the Matterhorn, Mont Blanc and, around the bend, Lake Geneva.

From this vantage point the glacier looks like an enfeebled monster withdrawing to its cave to die. Visitors arrive as sightseers, thirsty for a blast of Alpine air and a gulp of high-priced Alpine coffee – but leave as mourners. Elegiac thoughts hang in the air like fine spray. It comes as a shock to realise that the grey-green water leaching out of the glacier's base, as if from a wound, is made from

The boarded-up Belvedere, whose windows once looked out on ice.

snow that must have fallen onto these peaks longer ago than anyone alive can remember.

~

The fact that the three formative rivers of France, Germany and Italy began in the same glacial height is a pleasant coincidence. But there is a more urgent aspect to it, too, because this shared height *is* melting. Thanks to climate change, and the carbon habit of modern times, it is dissolving at a hectic pace: evaporating before the disbelieving eyes of those who come this way, who are stunned to see how far down the ice used to reach (marker posts provide the dates and distances).

This has frightening implications. The streams that dash through the rocks up in the mountains, sparkling in the sunshine, are fed by thawing snow and ice. And herein lies the instinctive genius of the natural world. The warmer the sun, the faster the glaciers release their husbanded store. In heatwaves they send fresh, chilled water into the rivers exactly when it is most needed; then, in winter, they preserve falling moisture as ice, lowering the risk of floods and refilling the deep freeze for next year.

The glaciers are adjustable. They are thermostats.

But these great castles of ice, which once seemed eternal, are dying. And this is not just an aesthetic matter, an icy canary alerting us to the loss of a shining view. It is a straightforward fact. The taps that have kept Europe's rivers full for thousands of years are running dry.

There remain a handful of determined objectors to this calamity, but it is no longer open to serious debate: the before-and-after images are incontrovertible. Modern sensors show that the Alpine ice has lost 10 per cent of its volume since the start of the century, and is in danger of losing *half* its present mass by the end of it if the world does not find a way to slow the overheating. And this is part of a bigger, grimmer picture. A 2023 survey by Europe's CryoSat satellite revealed that in the last decade Earth's 200,000 glaciers, situated mostly at the poles, had lost 2,720 *billion* tonnes of ice – enough not just to fill the rivers but to make them overflow.

There is little need to debate the cause: global warming is almost universally accepted as a dangerous fact of modern life. And average temperatures are rising even more sharply in Switzerland than elsewhere (by up to 2 per cent in recent years) thanks to a meteorological quirk caused by its high mountains. They have turned the Rhone Valley into a suntrap, with 300 days of blue sky every year, and tropical vegetation such as cacti that are found nowhere else on this latitude. While that might be lovely for tourists, it is not so good for ice. In early 2024 a petition brought by 2,000 members of the Club of Climate Seniors successfully took the Swiss government to the European Court of Human Rights, seeking a landmark ruling on the need for faster progress towards carbon neutrality.

The country was already aware of the problem: some of its (and the world's) most famous ski resorts had become increasingly dependent on snow cannons. Every single one of them was having a heated debate about its own future viability. The petition attracted wide support and won its day in court.

Nowhere is this issue better demonstrated than at the head of the Rhone, where the base of the glacier is now a grey pool with a grey river trickling out of grey rocks – robbing the once proud Belvedere of its purpose.

This is catastrophic decay. Until recently the Alpine glaciers felt permanent, timeless, part of the 'eternal' order. But the speed of their retreat forces us to reflect on their importance again – not as longstanding friends, but as founding fathers whose days may be numbered. We can no longer see them as immortal. And while they may have already accomplished their first great task – the million-year duty that carved out the valleys down which the rivers could run – they still perform an important function feeding and regulating the rivers that flow from their icy noses, and their disappearance would have consequences no one can confidently predict.

Of course, the rivers are not fed by glaciers alone, so this does not mean the waterways face imminent extinction too; the Rhine has 144 tributaries, the Po forty-four and the Rhone forty-three. And the climate change that is threatening the glaciers is if anything filling these streams with increased rainfall, in ever more violent bursts. So it would be premature to start writing their obituaries. Rainfall and glacial meltwater have been running down these slopes for longer than we can imagine, through tectonic collisions, ice ages, droughts and floods. The rivers' courses have shifted many times. But they have not died. The harm done by modern carbon-burning is extreme, to be sure, but the greater water cycle feels mighty enough to outlast our worst attempts at sabotage. It seems almost conceited to imagine that we could halt so vast a pattern: perhaps our modern crisis might turn out to be a transient episode in a longer story.

We can but hope.

~

The things we take for granted, John Betjeman wrote, are the things we'll miss the most. It's the kind of saying that can't help but nag at our consciences when we contemplate glaciers. Until recently they seemed other-worldly – too big to fail. Now they feel defeated and flimsy.

They may comprise only 10 per cent of Earth's surface (mostly at the poles) but they still hold two-thirds of the world's fresh water. If they all melted, the oceans would rise by 18 inches, enough to drown a frightening number of coastal cities.

There is a strange alchemy to ice. Stretched thin, it is as brittle as glass, but when it is thick it becomes 'plastic' (bendable) and is supremely strong. Because most of the glaciers' weight is in the centre, this is where they move fastest, drawn down by gravity and lubricated by water running through their cracks. On the surface ice appears white, but once the air has been squeezed out it becomes a radiant turquoise-blue.*

Glaciers tickle our senses in another way. Oozing down from the heights under the pressure of their own weight, their enormity makes us shiver. So much power, so lightly held. And they look tranquil. But that shining exterior, fresh as swans-down, is an illusion. Close up we can sense the titanic forces acting on and through them, and their formidable age. They are made of frozen water, but also frozen time. No wonder they have often been demonised as malevolent monsters. They seem half alive, yet their gigantic stillness spins us back to unimaginable pasts.

---

* The serious explanation for this is that the long (white) wavelengths of light are absorbed by compacted ice, while the short (blue) wavelengths are scattered; the more everyday version suggests that as the ice compacts the air bubbles are squeezed out, creating the same blue colour found in deep water – for much the same reason.

6

THE DEBACLE

On 5 April 1815, an earth-shaking upheaval on the other side of the world was preparing to make its presence felt. Mount Tambora, the highest volcano in Indonesia, was beginning to rumble and spit.

No one could suspect that this could possibly have any bearing on the history of Europe's rivers, but it did.

A series of detonations shook the island, and burning ash rained down from the darkening skies. That was only a cough. Five days later the whole mountain blew up. Three towers of flame jetted into the clouds and a glowing wave of boiling rock blasted out of the summit. It was accompanied by an ear-shattering eruption that sent 175 cubic kilometres* of rubble 30 miles up into the air – a colossal bang, said to have been ten times as fierce as the explosion at Pompeii – perhaps the loudest noise ever to rattle the earth.

Modern models estimate that the explosion was perhaps 10,000 times as powerful as the Hiroshima bomb (estimates vary: some say it was 50,000 times). At any rate, it shook windows and made people start in Batavia (now Jakarta), 800 miles away, and

---

* How big is 175 cubic kilometres? Not easy to say, since a lot of that cloud was composed of air. But it has been estimated that the material residue of Tambora was enough to coat the entire surface of Britain with a one-inch layer of ash.

was heard as a distant thump, like cannon fire, hundreds of miles further afield. In Java, the sound was so distinctive, and so loud, that the British colonial authority believed itself to be under attack, and urgently dispatched troops to investigate.

Up to a hundred thousand people died in the whirlwind triggered by the convulsion, though that may be a low guess. The whole region went dark; an entire culture was buried. Tambora can claim the horrible distinction of being the only natural disaster ever to have obliterated a language.

The previous day it had been – at 14,000 feet – almost as tall as Mont Blanc. Now, as the dust began to settle, the top half had disappeared altogether; there was nothing left but a crater.

And that dust did not settle. The plume flew up into the stratosphere, causing strange weather effects across the world. In nearby Borneo it fell as hot ash; in Italy it turned the snow red; New York suffered dirty fogs. Phenomena such as these – lurid sunsets and rosier-than-usual dawns – were harmless; but some of the repercussions were serious. So dense was the curtain of debris that the average global temperature fell by a degree, enough to cause ruinous damage. In 1816, crops failed in Europe and America, triggering the worst famine of the nineteenth century. It became known as 'the year without a summer'.

Today we know all too well how a calamity in one part of the world might cause a tremor in another. We call it the 'butterfly effect', and enjoy imagining the subtle chain reaction by which the flap of a wing in one hemisphere can produce a debacle on the other. But in 1816 this was absolutely not common knowledge. The weather was seen as an act of God – a force beyond human computation. Tambora might have been on another planet. It took months for news of its demise to reach *The Times*.

Two summers later, in a steep valley in Switzerland, above the Rhone, something else happened, something equally inexplicable.

A river dried up.

By rights, with spring on the way, the River Dranse should have been full of rushing water. But in the early months of 1818 the opposite happened: it slowed, shrank, dwindled, shrank a bit more, and by April had tapered away to nothing.

For the farmers and villagers in the Val de Bagnes there could have been no greater setback. Their home was a winding dale running from the Rhone up to the Great St Bernard Pass, where monastery dogs helped patrol the route over to Italy.

For the valley's rural inhabitants the water of the River Dranse meant . . . everything. Their crops, their animals (for milk, eggs, butter, cheese and meat) and their own daily lives depended on it. Scrambling up, they found that the snout of the Giétro Glacier had fallen across the channel and frozen solid. A tall ice dam was blocking the entire valley. And now a lake was forming behind that dam, which made the villagers even more fearful – because this had happened before. The folk memory of the region had not forgotten the great *inondation* of 1595, when a glacial pool had smashed through its retaining wall (which, being ice, was lighter and weaker than the water pressing against it) and swept down the gorge, sending 140 hapless souls to their deaths.

Those who lived in the valley were no strangers to the idea that ice advanced and retreated. That was an annual event. The *montagnards* knew from first-hand experience what glaciers could do. They had seen these mountain demons come and go with their own eyes, year after year, crashing their grim way through homes, barns, trees, walls, roads, chapels and anything else that stood in their path.

But the brainy scientists and learned theologians in Switzerland's elegant cities dismissed such talk as primitive rural superstition – surely *everyone* understood that those blinding-white glaciers were a magnificent part of God's handiwork, a flourish on this divine mountain scenery. The idea that they were surly forces of nature, with their own moods and habits . . . that was charming, but surely a childish whimsy.

It takes a conscious effort to realise that it was not generally understood, back then, that glaciers were active and dynamic, dragged down by their own immense weight and the plasticity of frozen water.

The first British expedition to the Alps, in 1741, had sent a group of young Grand Tourists to a hamlet named Chamonix, at the foot of Mont Blanc. The pamphlet describing their adventure was written later by William Windham, a friend of the eminent Pitt family, and published in 1744. It included an account of 'our journey to the *glacières*', and contained a revealing description of the crevasses in the ice: 'Our guides assured us that these cracks change continually, and that the whole *glacière* has a kind of motion.'

We should let that sink in. It was common knowledge to 'our guides' – the Alpine villagers – that *the whole glacière had a kind of motion.*

It was only the rest of the world that did not know.

~

One of the valley's senior citizens, Jean-Pierre Perraudin – a farmer, guide and civic leader – was especially concerned by the ice dam and lake. On his clamberings around the mountains he had made an extensive study of the local rock formations, and was convinced that the Alpine glaciers must once have touched . . . everything. The scrapes on the stones bore witness to the fact that this gorge must have been chiselled out by an enormous thrust of ice. And the granite boulders littering the valley – too heavy to have been tossed about by water alone – had clearly been flung from the heights.

He could not take the ice dam lightly. And he was also a man of action. Having served as a lieutenant in the Valaisanne army (Valais was not yet part of the Swiss Confederation) he had seen Napoleon's troops tramping up his peaceful valley en route to Italy. It must have been a remarkable sight, that column of 36,000 men marching beside the Dranse, with mule trains hauling cannons over the rocks – it reportedly took two weeks for the soldiers to pass. Now, as the frozen lake grew in front of Perraudin's eyes, something more worrying was coming into focus. A cascading river was an even more dangerous opponent than an invading army. As one of the most respected figures, aged fifty-one at the time, in a valley commune of more than twenty villages, it fell to him to sound the alarm.

The wider world was ignorant of it all until he alerted the authorities down in Martigny. So help, when it came, arrived in the unlikely and relatively junior form of a twenty-one-year-old canton engineer named Ignace Venetz. In due course, his discoveries would change the world. But first, Perraudin provided him with lodgings, and the two men began to talk.

Venetz was serious-minded, and quick to grasp the severity of the crisis. The frozen cone was not a danger, but the lake behind it was a palpable threat: over a mile long, and growing deeper by the day. In a letter to his supervisor he called the situation 'fearful', adding that the town of Martigny, where the Dranse met the Rhone, was in danger.

Something had to be done.

We need to keep reminding ourselves that no one alive at that time had the least idea what to do in such a crisis. Venetz was on his own.

Calculating that if he drilled a hole through the ice the water should stream through in a controlled manner, he drew up a plan and recruited a team of Italian labourers to do the work. The idea was to dig out the ice tunnel *above* the present waterline, so it could function as an overflow when the water level rose, which it certainly would.

Nothing like this had been tried before.

Time was pressing and the workers were unhappy. The hole needed to be 200 yards long, and it was freezing in there, with ice water up to their ankles and hands too frozen to hold or swing a pickaxe. And the surrounding ice groaned, as Venetz put it, 'like a wounded animal'.

After a week the labourers downed tools and walked away. It was left to the villagers to take up the ice picks.

In some later accounts those Italian labourers were depicted as cowards, but it is just as likely that they were experienced men of the mountains who understood the language of ice – its menacing sighs and grumbles – and knew that their tunnel was not safe. If so, they were right, because their own axe blows were fatally weakening it.

Aware of the danger, Venetz placed beacons up on the summits to act as a warning system in case the worst should happen.

Which it soon did.

On 16 June an awful roar echoed across the valley – exactly what the villagers were fearing. Venetz urged everyone within earshot to race uphill as fast as they could, and then did the same thing himself.

By the time Venetz and Perraudin were racing for the high ground, the catastrophe in the Val de Bagnes was already unfolding. At 4.30 in the afternoon the ice wall broke and released the lake, roaring down to the Rhone at Martigny, where a nine-foot wave sloshed through the town. Plucking up trees as battering rams, it knocked over everything that stood in its path. Jean-Pierre Perraudin's chalet, more solid than most, was swept away, and the village of Champsec was obliterated. A plaque on the rebuilt church recalls that dreadful afternoon for modern passers-by.

In all, three hundred buildings perished, and seventeen bridges. Some of the casualties – such as the eighty-six-year-old Marie Troillet – fell because they were too slow. Others, like the infant Sophie Mermoud, eight days old, died because they were too small: she fell from her basket in the rush to escape.

There was no time even to light the beacons. Although the death toll of thirty-six was lighter than in 1595, it left Venetz with a new-found respect for the daunting power of Alpine ice. Indeed he came to believe that this region of the Alps must once have been buried under a colossal ice sheet, put his findings in a paper, and in 1834 presented them to the Swiss Society of Natural Sciences in Lucerne.

The audience was unmoved. Everyone knew that the boulders in the valley had been put there by Noah's Flood – it would take more than the observations of a farmer and a local official to persuade them otherwise.*

But one man, a mining engineer named Jean de Charpentier, was intrigued. As it happened, he had met Jean-Pierre Perraudin three years earlier and had stayed in his house, so he was not unfamiliar with his theories. He decided to head back to Lourtier and look up his old acquaintance. This time he conducted a detailed survey, wrote up his thoughts and presented them to the Helvetic Society in Lucerne.

---

* In his famous *Principles of Geology* (1833), the British geologist Sir Charles Lyell admitted that these so-called 'erratics' *might* have been transported by icebergs. But even he did not go so far as to suspect that glaciers might have been involved.

Once again, the audience was not much interested. But at least one of its members was open-minded. Louis Agassiz, Professor of Natural History at the University of Neuchâtel, had been a pupil of de Charpentier, and though he wasn't entirely convinced, he was polite enough to accept an invitation to come to the mountains and take a closer look.

He pushed rows of sticks into the ice to measure its speed, had himself lowered into a crevasse to inspect the ice's inner workings, looked at the scratches in the rock and the stones strewn across the ice, and suddenly understood: this landscape clearly *had* been sculpted by frozen rivers. And the Rhone Valley could hardly be alone: the whole of northern Europe must once have lain under an unimaginable sheet of ice.

This time there was a flicker of interest, so Agassiz went to the Unteraar Glacier, higher up the Rhone Valley, and compiled further evidence. He tracked the ice's thickness, proved that the middle flowed faster than the edges (and the surface faster than the base), studied the alternating lines of blue and white, and invited other scientists to inspect his techniques. One of his visitors, the Oxford Professor of Geology William Buckland, amused his hosts by wearing his academic robes out on the glacier. And even though he was a devout Christian theologian (later Dean of Westminster), he supported Agassiz's findings.

Anyone still tempted to resist the thesis had to admit defeat in 1863, when three ragged corpses were found at the foot of the Glacier des Bossons in Chamonix. An elderly guide in the town, which by then had become a famous resort, was able to identify the frozen bodies as having been his luckless companions on an expedition to climb Mont Blanc back in 1820 – close to the time when Perraudin and Venetz were climbing up their dried-up riverbed to squint at the ice. The 1820 ascent was a disaster: the three guides were pushed off the slope by an avalanche and lost in the crevasses on the Bossons Glacier, never to be seen again . . . until they were spat out, recognisably themselves, forty-three years later.*

The glacier brought them home, at a speed of approximately 164 feet per year. Who now would dare say that ice was inert?

There is a moral here. Intellectual advances are not always – and perhaps not often – achieved by academic theorists performing abstruse calculations, but by ordinary people with a close knowledge of their surroundings and a large supply of natural curiosity. Just as the inventive spirits who sparked Britain's Industrial Revolution were not university scientists but skilled craftsmen seeking solutions to routine quandaries, so it was Perraudin's deer-hunting scrambles that changed the world.

It all means that one of glaciology's founding fathers was a chamois hunter in a remote Alpine valley. As so often happens, he is not much lauded. It was Agassiz who took the idea to America, became a star on the lecture circuit, and ended up giving his name to a mountain in the Bernese Oberland (the Agassizhorn), a lake in South Dakota, and a newspaper in British Columbia (the *Agassiz-Harrison Observer*). Perraudin, the 'country fellow from Lourtier', is mostly overlooked.

It has fallen to Marguerite Perraudin, the woman who married his last male descendant, to honour his memory. In 1993 she turned his house into a shrine to his powers of observation: the Maison des Glaciers. It stands there still, on the Chemin Jean-Pierre Perraudin.

---

* One of the dead guides was Pierre Balmat, brother of Jacques Balmat, the Chamonix pathfinder who conquered Mont Blanc and is commemorated by a statue in front of the casino, his finger pointing excitedly at the summit he brought to heel.

The Perraudin house in Lourtier, now a museum of glaciation.

The exhibit includes a white paper model of the glacier that did so much damage. At first it seems innocuous. Then, unnervingly . . . it moves. And as it scrolls imperceptibly forward, the helpful maps, images, videos and letters on the wall spring to life, coming together to form an eloquent memorial to that dreadful but significant day. It is an evocative reminder of the extent to which science can evolve in strange ways and surprising places – and how the flowing ice has shaped not just the geography and history of the Alpine region, but gave fresh energy to new fields of scientific inquiry.

~

The discovery of flowing ice was not the only scientific breakthrough to take place in the balconies above the Rhone. The ice contained further secrets. One of the mysteries in nineteenth-century Switzerland, for example, was why so many of its citizens had the unsightly neck bulge known as a goitre. A 1921 survey carried out by Otto Bayard, a doctor who lived near Zermatt, revealed that 70 per cent of Berne's children had a swollen thyroid, and a third of nineteen-year-olds were similarly affected. Some assumed it was a genetic defect; a few extremists even whispered about 'racial hygiene'.

Why was God punishing the Swiss?

Heinrich Hunziker, a medic and poet who lived on Lake Zurich, speculated that the swellings were caused not by anything invasive but by the precise opposite – an absence. He had written a book on the subject in 1914, building on the work of a French chemist who had written a paper on iodine deficiency back in 1852.

When Bayard came across Hunziker's work, he wondered whether the goitres might be eased by iodine, a trusted cure-all for many ailments, and he trialled a cure in the neighbouring village of Grächen, where 75 per cent of the children had the enlarged thyroid. Using a snow shovel to stir gallons of iodine into salt, he administered modest dosages to the village's many sufferers.

Their goitres vanished almost overnight. Iodine *was* the magic bullet.

Experts struggled to believe that so grave a problem could have such a simple cure. But the evidence was unambiguous. And the explanation, once again, lay in physical geography. The depth of the ancient ice sheet, and the grinding motion of Perraudin's glaciers, had torn the topsoil away from the Alpine region, and that was where iodine, which occurred naturally in seawater, was stored. Switzerland was denuded: there was not enough iodine in the crops, meat, milk, air or water. The people were iodine-deprived. There was nothing more to it than that.

A federal commission recommended using Bayard's cure, and table salt was impregnated with iodine so that it could be sprinkled into the national food chain as a preventative exercise in public health. The result was miraculous. Six months later, Switzerland was goitre-free. The rate of deaf-mute births fell from one in 600 to almost nothing.

Thus was it discovered that one of the side-effects of glaciation was to scrape iodine out of the landscape.

This story was well researched and presented by Jonah Goodman in the *London Review of Books* in November 2023. It was, he wrote, 'the first attempt by a government to improve the lives of an entire population by adding a chemical to its food supply . . . One of the most successful public health measures ever devised.'

But one thing Goodman did not mention was the importance of the Alpine backdrop itself: the mountainous Rhone Valley was not simply a delightful canvas for these discoveries about iodine, but their architect. Once again it was Alpine ice and flowing water that built the foundations of the human societies through which it pushed and flowed.

7

MOUNTAINS AND MONSTERS

The eruption of Mount Tambora did a great deal to advance the infant sciences of glaciology and climatology. But those were not its only influences in these Alpine mountains.

As chance would have it, after visiting the Maison des Glaciers in Lourtier, I went down to Martigny, the Roman settlement that was flattened on that terrible day. Originally built on the sharp bend in the Rhone in order to control the Great St Bernard Pass, it is now a pleasant town surrounded by vines. And its stylish art gallery happened to be showing Turner watercolours.

As I walked in, I noticed something obvious. Those fiery sunsets for which Turner was famous, along with his yellow skies and thunderous cloudbursts . . . it suddenly seemed likely that these too could have been inspired by the surreal weather that caused the 1816 disaster.

In the summer of 1816 Turner was touring through Yorkshire and Lancashire (the vermilion sky in *Lancaster Sands* suggests there were bravura sunsets there, too) but the following year he took his watercolour box across the Channel and the after-effects of the 'year without a summer', caused by the eruption, are visible in most of his skies.

He was not the only visitor from the world of the arts. The writer Percy Bysshe Shelley went to Chamonix to see Mont Blanc and was struck by the way the 'ice-gulfs' flowed from the peaks like living things: 'the glaciers creep / Like snakes that watch their prey'. And he and his new wife Mary* recorded in their joint journal (published anonymously in 1817 as *History of a Six Weeks' Tour*) that 'these glaciers flow perpetually into the valley ... driven onward by the irresistible stream of solid ice'. Later Mary would call them 'tremendous and ever-moving'.

Lord Byron and his friend John Polidori took over the Villa Deodati on Lake Geneva, with the Shelleys close by. Thanks to the weather they made few excursions. Mary Shelley's precocious journal – she was only eighteen – referred to 'incessant rain'.

In a famous anecdote, Byron proposed that they while away the gloomy days by telling ghost stories – with interesting results. Mary Shelley brooded on the heresy that it might be possible to 'galvanise' a corpse and came up with *Frankenstein*.† In creating her monster she was in part inspired by the physician Erasmus Darwin (Charles's grandfather) whose experiments suggested it might be possible to breathe life into vermicelli, but she was also influenced by a place she had passed on the Rhine. It is not proven, but when her party

---

* Their romance began, scandalously, two years earlier when she was sixteen and Shelley was an already married twenty-one-year-old. They met, in secret, at the gravestone of her mother, Mary Wollstonecraft, whose stirring writings on 'the rights of women' did not discourage her daughter from eloping to the mountains with her poet lover.

† Galvanism was not just a word. It was Luigi Galvani, an eighteenth-century Italian scientist, who had made the legs of a dead frog twitch by passing electricity through them, giving rise to hopes that life could be recreated in a laboratory.

stopped at Gernsheim, near Darmstadt, she might well have heard talk of a ruined tower 10 miles from the river named Frankenstein Castle. It was famous mostly as the birthplace of a seventeenth-century alchemist, Johann Dippel, who had invented a healing oil he called the 'elixir of life', and who (it was whispered) had disinterred real corpses for his strange experiments.

It may be nothing more than a coincidence. But if Mary Shelley did *not* hear about this memorable character, and came up with the name for her doctor hero on a pure whim, well . . .

Her fellow guest John Polidori came up with something strikingly similar and of its time by inventing an equally famous figure: Dracula.*

In time his story, *The Vampyre*, would be superseded by Bram Stoker's much bolder characterisation, but it turned out that he and Mary Shelley had given birth not just to timeless characters but to a shocking new genre. And the fact that both were forged in the gloom of the same freak weather is telling. It means that the Tambora-wrecked summer of 1816 can claim not just to have laid down the conditions – stormy skies and cold, black afternoons – that gave rise to new fields of scientific inquiry, but also to have played midwife to a sensational new note in English literature: Gothic horror.

Who would have thought that an Indonesian volcano could give birth to a potent literary form in a faraway continent?

---

* Polidori was the son of an Italian émigré who settled in London as a scholar and befriended the Pre-Raphaelite group of artists and poets when his sister married Dante Gabriel Rossetti. A medical student at Edinburgh, he was also Byron's doctor.

8

DAM AND BLAST IT

There's an unusual feature in the valleys above the Rhone in Switzerland. The road that winds up into the mountains from Sion, the regional capital, runs towards Italy through pine woods, and at first there is nothing to suggest that it won't end, like most such routes, in a heart-stopping Alpine view. But this one has a different climax. After zigging and zagging along steep cliff edges, the road narrows, slips past a couple of precarious villages and bursts into high pastureland.

At a certain point the valley divides. The left branch, the Val d'Hérens, climbs into the snowy heights of the Matterhorn range. The right-hand road leads through lush fields full of warrior cows until something strange appears: a blank grey wall that, the closer one comes to it, seems to blot out half the sky.

Towering above the dark pines, it looks like a concrete silo – a missile base, perhaps, or the ostentatious lair of a Bond villain. In fact it is the retaining wall of Europe's highest dam: the Grande Dixence. It is sixty years old, but it has a longer history than that – one that tells the story of how Switzerland's rivers became the source of its energy.

As early as 1882, a Swiss engineer named Théodore Turrettini was putting turbines into the river at Coulouvrenière, near Geneva, creating enough power to drive the local textile industry. People had been dipping wheels into water since Roman times, but it was becoming increasingly apparent that the future was going to be driven by electricity (Edison founded his Electric Light Company in 1878). The ability to extract power on a large scale was now a matter of prime importance – and Turrettini had shown Switzerland an exciting new future built on water.

It was at the end of the First World War that Switzerland, alarmed by the wartime realisation that its economy depended on German coal, began to look seriously to its mountains for a more reliable source of power. It was blessed with Europe's most potent hydroelectric assets, and the Rhone Valley, which contained more than half of the country's glaciers, was its highest point. The river in the Hérémence valley above Sion – the Dixence – was of no great size, but in 1927 the national energy company bought the licence to contain the small Lac du Dix with a huge dam and set about creating the necessary infrastructure.

It was not a simple matter of barricading the stream and waiting for the bowl to fill. The plan involved the construction of four pumping stations to funnel meltwater from the glaciers into the new reservoir. That meant building 60 miles of tunnels as well as the colossal retaining wall, and it took a 1,200-strong army of labourers to deliver the project. Nothing like it had been tried before, but by 1935 the dam was spinning turbines at the Andoline power station down in Sion, and the future of energy – long before the wider world was fretting about sustainable sources – was hoving into view.

After the Second World War an even more ambitious plan was hatched – to enlarge the lake by building an even bigger wall half a mile further down the valley. This one required the installation of a spider's web of new cableways to hoist over 200 million cubic feet of concrete up to the site, where half-a-dozen mixers could make it ready for pouring at a rate of 800 tonnes per day. It was a Herculean task and, once again, the result was ground-breaking: at nearly 935 feet (almost the height of London's Shard) this was instantly the tallest dam in the world.*

After generating enough power for 400,000 households, the water released by this engineering wonder entered the Rhone at Sion before flushing through Lake Geneva to the south of France. Though rivers usually have a single officially recognised source, they have a thousand feeder streams, and the one that flowed out of the Grande Dixence was unusually well behaved and useful.

Folded into this large story was an intriguing footnote. One of the junior workers on the Dixence project was a Swiss-French student who had grown up in nearby Nyon, on Lake Geneva. While the Grande Dixence was being planned, he had been struggling to make his way in Paris, where his grandfather had founded Banque Paribas. He had failed his *baccalauréat*, which ruled out a career in the family bank, but he did not greatly care, because he aspired to a future in the arts and was hanging around the trendy clubs, cafes and artistic haunts of Paris's Latin Quarter, hoping to develop the personal contacts on which such a flighty career might depend.

A job on an Alpine building site was not part of this dream. But his mother's partner was involved in the Grande Dixence and offered him work as a telephone switchboard operator. He was immediately impressed by the cinematic potential of this lofty construction

---

* A title it held for more than a decade. In 1972 the Nurek Dam in Tajikistan beat it by 50 feet. To this day that remains the world's tallest.

site. Aiming a borrowed 35mm camera at the dam, he wrote a terse script and delivered it himself, backed by high-minded music: Bach's Double Violin Concerto.

This is how the noted *cinéaste* Jean-Luc Godard came to make his first picture. In this he was mimicking Michelangelo Antonioni, whose own screen debut *Gente del Po* (*People of the Po Valley*) depicted the hard lives of the bargees who ferried agricultural produce along the river. That was filmed in 1943, and the way it looked at the families who kept the barges going ('a grim life that never changes') or the farmers in their unstable reed huts, brought a novel note of realism to Italian cinema.

Godard became famous for saying that the only thing he needed to make a film was a girl and a gun. But his first venture was this Grande Dixence documentary. There were hints of poetry: the cement bucket dangling from the high cable was named Blondin, after the high-wire artiste, and the tangled machinery was described as 'an organism of iron and steel'. The sixteen-minute film (*Opération Béton*) became a small but notable bit of cinema history.

~

The Grande Dixence is a site, but also a sight – a stony aquarium held up by the power of human ingenuity. Like so many things in Switzerland, it doubles up as a tourist attraction: there's a zipwire for thrill seekers. But it is not unique. On the contrary, it is one of 200 such colossi. And those are only the heavyweights in these hills: there are a thousand smaller water tanks, all maintained by the local cantons. In 1924 a barrage was built at Chancy-Pougny, on the Swiss–French border; another appeared at nearby Verbois during the Second World War; and a fourth, the HEP plant at

Holding back the water: the Grande Dixence dam, high above the Rhone at Sion.

Le Seujet, in 1995, meant that the Rhone was giving up its energy four times over. That was enough to supply a significant portion of Geneva's needs. Switzerland is now by some way a world leader in retained water: the most dammed country on Earth.

Switzerland soon discovered that its new concrete water tanks gave it a powerful edge in the energy-hungry modern world. The Alpine dams allowed it to develop a hydroelectric energy sector big enough to fuel the national need for heat and light. In 1970, an unusually high proportion of the nation's energy came from hydroelectric plants. And quite apart from being environmentally friendly, this proved to be a timely asset in a world about to be reshaped by a series of oil shocks.

In the mid-1970s, the oil-producing states (led by Iran) banded together to squeeze the supply, and the price of oil multiplied from three dollars a barrel to more than a hundred, triggering economic panic across the developed world. Switzerland, however, was noticeably more self-reliant than its European neighbours: its thousands of hydroelectric plants generated nearly a quarter of its energy needs. Although it was not completely immune to the crisis, which triggered one of the largest transfers of wealth the world had ever seen – from the West to the OPEC oil producers – sensible Switzerland, secure in its lakes and mountains, had an awesome natural resource: the water tumbling from its hills. Swiss motorists had to pay more for petrol, like everyone else, but the national grid, in a land of icy winters, was not so badly affected as the networks of its competitors.

It is hard to gauge how much of the country's modern prosperity derives from this fortunate fact. But the Rhone still supports 150 hydro-power plants, and Switzerland's national railway is 100 per cent electric. The resilience that its waters provide, in the context of today's heightened anxiety about carbon, is clear.

However, this does introduce a quandary. The rising temperatures responsible for dissolving the world's glaciers are also threatening the dams, since it is diminishing the runoff from the melting ice on which they depend. The implications of this are not mysterious. Yes, increased rainfall will keep the rivers alive. But without the natural modulator that stores that rain as ice, the glaciers and the dams they feed will be less reliable. And the power plants attached to the dams, which once offered liberation from the polluting geopolitics of fossil fuel, now find themselves endangered by the global warming they were built to counter. If the reservoirs do run dry, those turbines may stop.

This is why futurologists are warning that the most dangerous global conflicts of the future will probably be ignited not by oil but by water. Serious scientists are quick to point out that the world is not running dry. But fresh water is extremely precious: 97 per cent of the world's water is salted, and two-thirds of the remainder is polluted, so just 1 per cent is all we have to squabble over, and the global distribution of that water is very uneven. The Middle East, for example, has 5 per cent of the world's people, but only 1 per cent of its water. In some regions the supply of water can become a flashpoint; India's proposal to deprive Pakistan of Himalayan water in the spring of 2024 (in response to a terrorist attack in Kashmir) was only the latest in a growing list of such threats. Repeated around the world, that imbalance – the friction it is bound to provoke and the compromises it will demand – may be the most urgent issue the world will have to face in the already foreseeable future.

9

HOTEL DU RHONE

The top of the Rhone Valley is a boulder-strewn meadow pock-marked with hollows and bogs. The glacier is no longer visible – it has withdrawn over its high stone lip, leaving only polished rock – but there are angry traces of its passing. A sense of wreckage pervades the scene. Stunted trees poke above the debris that has fallen off the bleak heights above.

A well-signposted path runs alongside the river – only a few feet wide at this point. Marker posts reveal the points reached by the historic glacier on various dates, encouraging hikers to imagine the valley when it was full of ice the height of a house. In 1856 it reached all the way to the bottom of the pass, to the grand hotel and railway station. Now it is a scrubby marsh, rewilding but crushed. Information boards call it 'an alluvial zone of national importance', but it feels more like the scene of a grisly battle, scarred with skeletal remains.

At this point the river is a missile, too violent to be either useful or a visitor attraction. It doesn't feel stable – a minor collapse would tip a lake-sized flood of freezing water onto this flood plain. The footpath gives the river a wide berth, and wooden stakes warn passers-by not to paddle: the stream is strong enough to pin even an adult swimmer to the ice-cold rocks.

One of those signs points out that from here it is 380 miles to Marseille, a jarring thought – it is hard to square the idea of that sweltering port with these waterlogged acres and mountainous cliffs. Yet this river is what those two worlds have in common.

Once its first steep descent is accomplished, however, the valley of the Upper Rhone mellows into a broad green glen, with a strong enough snow record to make it a notable centre for cross-country skiing. The walls on both sides are high, so although the villages are small – just groups of farms clinging to the rare spots of solid ground – there is a steady stream of cars and trains. Not many people live here, but the crowd that passes through is substantial. Two cable-car stations hoist people up from Martigny and Sion, thirsty for grand views and sparkling air, to the Aletsch Glacier – Europe's longest ice flow, and a spectacular provider of tributary water to the onrushing Rhone.

There is plenty of nature in this extraordinary scenery, but there is also nurture. The flow of visitors passing through this landscape shaped one of Switzerland's most renowned national characteristics: its tradition of hospitality.

~

In the village of Blitzingen, down in the valley, lived a young man called Alexander Seiler, who grew up working in the fields like everyone else. But he was given an introduction to the hotel trade by watching his two sisters hard at work in the Blitzingen Gasthaus. So, in 1841, when his brother – a chaplain in Zermatt – urged him to come and join him up the mountains, he went. He was quick to see the potential.

Zermatt was home to a new pastime – mountaineering – and the number of visitors was on the rise. Seiler saw his opportunity, rented a chalet and turned it into a thirty-five-bed hotel, the Monte Rosa. In due course he bought it outright and became a pillar of the community. But he didn't stop there. In 1857 he expanded by building the Grand Hotel du Glacier du Rhone in Gletsch at the foot of the Furka Pass. It was an instant success. Guests would line the balcony to inspect the famous glacier above their heads – one of the great sights of Europe: a wall of slow-moving ice with a famous river trickling out of its snout.

Buoyed by these successes, in 1867 Seiler bought one of Zermatt's smartest properties, the Mont Cervin. And again his timing was good. Two years earlier the English climber/artist Edward Whymper had conquered the Matterhorn, adding to the mountain's terrifying reputation by losing over half his party on the way down (four of the seven-man team fell to their deaths). The disaster made headlines across Europe, and Zermatt became the epitome of Alpine adventure. Visitor numbers swelled.

That made Seiler a power in both the town and the canton. And he was not finished. From the hotel in Gletsch he could see the bend in the road on the Furka Pass, with its close-up view of the glacier, and conceived the idea of building a new hotel up there for his son. This was the handsome pile of *belle-époque* glamour we have already seen – the Belvedere. It opened in 1882 and, like all of his projects, was filled with mod cons. *Electric light! Taps! Lifts! Flush lavatories!* Seiler was not doing anything radical, merely keeping up with the latest trends. And as a leading member of the National Council, Seiler was able to rubber-stamp the construction of a road over the Grimsel Pass (to Berne), which only increased the number of travellers flowing past the hotel's front door.

In 1884 he opened another resort hotel, the Riffelalp, on a sunny shelf with views of the Matterhorn. At first it was accessible only

Wish you were here. The hotels clustered around the glacier in this 1895 postcard helped establish the Swiss tradition of hospitality.

on foot or by mule, but progress was moving fast, and in 1889 the Riffelalp became a station on the cog-and-pinion mountain railway, served by pretty red trains. It was the highest guest house in Europe. And it is still there today, full of ramblers in the summer months and skiers in the winter.

Seiler himself died before the opening of the railway line from Visp to Zermatt, which promised to hoist even more passengers from the Rhone to the snowline. In the event, his coffin was ceremoniously taken down to be buried in Brig on the train's inaugural run.

He was not the only hotelier to emerge from these parts. The village of Niederwald, only a mile or two away on the high ground above the same river, was the birthplace of an even more celebrated host: César Ritz. Born in 1850, at the age of twelve he was sent to the Jesuit school in Sion, and there began a varied career that carried the spirit of Swiss hospitality across Europe to Nice, Paris, London and beyond, through a chain of hotels that made his name a byword for luxury. Ritz himself suffered a nervous collapse in 1913 and went back to the Rhone Valley, to a clinic in Lausanne on Lake Geneva. When he died four years later he was buried in the village where he was born and, as its most successful son, is commemorated by a statue, a railway station and a birthplace museum that contains a reproduction bedroom from the showpiece Paris hotel.

Davos ... Gstaad ... Interlaken ... Lugano ... Lucerne ... Montreux ... Murren ... St Moritz ... Zermatt. These and other fashionable resorts turned a country that until then had seemed quaint and behind the times into a state-of-the-art playground. Between 1888 and 1914 (when the First World War put an end to such leisure trips) the number of 'Grand' or 'Palace' hotels more than doubled from 1,700 to 3,500. In the Canton of Valais alone, the number of hotels jumped from 79 to 320. The coming of the railway was rapidly making Alpine scenery a prime attraction for European travellers, and since the only people who *could* travel were well-heeled, the hotels founded to supply their needs were opulent. The Mont Cervin in Zermatt continues to offer 'unobtrusive luxury' to anyone willing to spend £776 a night for a room with a mountain view.

The luxury was new, but the hoteliers were building on a venerable tradition that was already centuries old. The pathways leading up to the Alpine passes, which naturally followed the rivers, were full of travellers – soldiers, monks and pedlars – even in Roman and medieval times, and they faced the constant threat of avalanches, mudslides, rockfalls and ice. The shifting weather was dangerous in itself: when the clouds blew in, visibility could shrink to zero, and the drop on all sides made it perilous to take so much as a single step.* There was an urgent need for hospitality in even the remote and unfriendly heights.

There were few habitable spots. But medieval monks, eager to demonstrate God's kindness, built sturdy lodges where travellers could count on finding straw beds, stables, log fires and, on request, supper. Thus was born the Swiss mountain refuge, or *hospiz*. The one at the Great St Bernard, the earliest of these havens, was built by St Bernard of Aosta before William the Conqueror set sail for England.† The *hospiz* on the Grimsel, on the route from Berne and

---

* It remains a high-risk place. In 2000 a mudslide at Gondo, where Kaspar Stockalper built his customs tower (see next chapter), killed thirteen people hoping to enjoy the views from the top.

† The famous dogs were a later, seventeenth-century, addition, and it seems that the brandy flasks on the knick-knacks also owe as much to fireside tale-telling as to actual fact. In Dickens's *Little Dorrit* (1849), the Dorrit family, freed from debtors' prison and newly rich, spend a night in the Hospiz St Bernard, gratefully seeing it as 'another Ark', its 'rough convent walls' a cheerful haven in this blasted wilderness.

with a view of the Rhone Glacier, is almost as ancient, dating to the twelfth century.*

The monastic refuges tied together a network of villages that lay on lanes once known only to monks and smugglers. But thanks to these sanctuaries they started to become well-trodden routes.

In the age of the Grand Tour, travellers came from the cities of the north, following the trails laid down by the rivers to see the classical wonders they had till then encountered only in libraries. The Alpine refuges in which they stayed were the forerunners of the grand hotels that came later. And since the word *hospiz* derives from the Latin *hospes*, meaning a stranger, foreigner or guest, there is more to this than a play on words.

The shape of the Rhone Valley arranged things such that it inevitably became a cradle of hospitality. Villages including Blitzingen and Niederwald lay on the only road from Geneva to Milan; a few miles above them the path forks – one way up to the Nufenen Pass and on to Ticino, the other past the Rhone Glacier at Gletsch to Andermatt and the Saint-Gotthard. They were positioned perfectly to cater to the traffic tramping up and down the valley.

It was Alpine ice that made the rivers that watered the fields that supported human life; the same ice also carved out the passes through the summits, drawing outsiders to this part of the world and creating a demand for board and lodging that Switzerland was more than happy to meet.

---

* Like the one at St Christoph, above St Anton in Arlberg, and many others, it is now a luxury hotel.

10

KING OF THE MOUNTAINS

Brig, the town where Seiler was buried, is so-called because it is indeed a bridge over the Rhone, and thus a transport centre. Like Blitzingen, it stands on the route up and down the Rhone, but is better placed in that it also commands the road up to the mountains. That makes it both an interchange for freight trains running through the Alps and a transfer station for the tourist trade. The *Bahnhof* resembles a palace hotel, and feels somewhat too large for a town of this size. But most of the traffic is not local: Brig is on the main line from Basel and Geneva, which brings in skiers heading for Zermatt and Saas-Fee, or sightseers en route to the Aletsch Glacier.

Nearly 6 million passengers a year pass through this little town, an extraordinary crowd for a place that in most respects feels out of the way. It could not avoid being a staging post. But its present prominence owes much to a long-dead character called Kaspar Stockalper.

Stockalper was born in 1609 – the year, for reference, when Shakespeare sketched out *The Tempest*, and the high winds of Reformation and Renaissance were still gusting across Europe. He was not an aristocrat; his father was a notary. But he became fluent in half-a-dozen languages (Latin, Greek, German, French, Italian and Spanish) as well as proficient in mathematics and law while attending the Jesuit College in Freiburg. Eager to broaden his horizons, he went on to study in France, Burgundy and the Low Countries before returning to Brig as well connected as he was educated.

Three pieces of good fortune helped him to prominence. The first was that he was born into a successful family – there had been Stockalpers in the valley since 1399. The second was that they lived in Brig – and he was canny enough to spot its potential. And the third spoonful of luck was that in the spring of 1634 he was asked to escort a French princess, Marie Marguerite de Bourbon-Soissons (wife of the Duke of Savoy), and her entourage of fifty courtiers over the hair-raising Simplon Pass, one of the few possible routes to Italy. In those days the passage was thought too dangerous for all but brigands: even Rome had not managed to build a genuinely secure road up there. But after assembling an escort of 200 men and 150 horses, Stockalper successfully steered the princess and her party over the mountains and on down to Milan.

It changed his life. He immediately became famous in the courts of Europe as *the* fellow for such enterprises.

A less alert man might have been content to count his blessings and bump up his fee. But Stockalper had now seen the true commercial value of Brig's location. He improved the mule track on the pass and set up a private horse-drawn service – using a trademark mustard-yellow postbus* – to run the crossing on a businesslike basis. Initially it was only a mail service – useful, but small. But it soon became a people carrier, and then a serious freight operator.

---

* The Simplon remains a busy trade route to this day, and the local bus is the exact same shade of yellow, in Stockalper's memory.

Stockalper found himself the sole holder of a lucrative monopoly: the distribution system for leather, amber, resin, turpentine and snails. In due course he would broaden his range to include iron, cheese, rice, silk, salt, livestock and other essentials.* By then he was no longer ferrying the loads himself – as master of the pass he was in a position to levy tolls for the use of the road he now controlled.

Ironically, the Simplon was one of the passes that did *not* have one of the old monastic refuges (until Napoleon ordered one to be built in 1801). But Stockalper turned even this to his advantage, since it made the Simplon especially lawless and thus boosted the prestige of its vigilant guardian.

With his new-found wealth he built a castle in Brig – a baroque folly, since it was never an actual fortress – whose fine courtyards and gardens dominate the town. The assembly hall has murals showing Alpine scenes, as well as a chapel and a courtroom. From the start, the castle was meant to be an administrative headquarters as much as a residence. And it had a symbolic function too – its three dome-capped towers symbolised the three wise men, the famous early travellers better known as the Magi.

It didn't hurt that one of them was named Kaspar.

The Simplon Pass soon had a tall, stone customs post: the Stockalper Tower. And when the Thirty Years War sent the armies of half-a-dozen powers marauding across Europe, Stockalper was able to charge all of them for the use of his pass. Indeed he was so synonymous with this trail that in Versailles he became known as *Le Roi du Simplon*. The sobriquet hints at something more than power, wealth and influence. Through his profitable willingness to

---

* The most precious of these was salt. Landlocked Valais had none of its own, yet had an intense need to cure and preserve food through its severe winters. As a result it imported some 750 tonnes per year, a sweet deal for the man who held the monopoly.

A tower built on tolls: the Stockalper castle in Brig.

deal with all sides, Stockalper was displaying what in due course would be Switzerland's characteristic neutrality. The Confederation was not a combatant in the Thirty Years War, though as many as 30,000 of its citizens fought as mercenaries for the rival armies. As a trusted intermediary, however, Stockalper saw that his own interests were best served by finding ways to deal with Austria, France, Spain, Sweden *and* Russia.

He even supplied some of the troops-for-hire himself.

Once again, the nature and shape of the Alpine Rhone was influencing the story of Europe. It allowed Stockalper to demonstrate that the fastest road to riches was peace. Europe was happy for Switzerland to be neutral, because none of its leaders could afford to be barred from those Alpine roads. And the whole of Switzerland, effectively, became one giant Stockalper Tower, a tollbooth in the mountains, exploiting the natural advantages of its position at the head of Europe's rivers and the foot of its Alpine passes.

~

Stockalper's most ambitious scheme never came to fruition. As a haulage operator he understood the importance of infrastructure, and planned a canal alongside the Rhone that would use the river to create a navigable road to Geneva. But by then he was a wealthy member of the Valais elite, a political leader and the father of fourteen children – the fact that the canal was never completed may have seemed a minor issue. And even without it he had contributed something new to a valley that until then was little more than a straggle of farms, snowbound for half the year.

The Simplon Pass, which connected Brig to Domodossola, would go on to become one of the most heavily used trade routes in Europe. Many years later, Napoleon's engineers would render it reliable enough to carry artillery, and though the emperor himself never came this way, this exercise did make it safe for motor vehicles when the time came.*

The road could not be relied on in winter, however, so at the end of the nineteenth century Switzerland's engineers followed up the success of the Saint-Gotthard Tunnel by drilling an underground route through this mountain too. Initially a joint venture between Switzerland and Italy (a treaty was signed in 1895), it was the former that provided most of the finance and expertise, and in 1903 it became a wholly Swiss enterprise.

The first train chuffed through in 1906, and a few years later the Orient Express was using it to steam luxuriously all the way to Venice. Soon its trains were being redesigned to carry cars, and Brig became a year-round shunting yard, especially busy in the frozen winter months.

Today, according to the Swiss Federal Office of Transport (FOT) some 40 million tonnes and a million lorries pound through the Swiss mountains every year, and a sizeable slice of that is carried through the Simplon: close to 10 million tonnes of freight, along with 5 million people, on 100,000 trucks and countless trains. The transalpine pass to which Stockalper played midwife has become a major pipeline for European goods.

Stockalper did not discover the pass – that had been done centuries earlier by Roman emissaries and the monks of the Middle Ages. Nor can he claim credit for the landscape over which he held sway – for that he had the rivers and glaciers to thank. But in a way

---

* A monument to Napoleon's road was built in 2005 to mark the 200th anniversary. It is one of two memorials on the pass: a 30-foot-high eagle in local stone made by the 11th Alpine Brigade was unveiled in 1944 as a tribute to the spirited Swiss people.

typical of Europe in the expansionist seventeenth century, he was the one who understood its potential. By dominating the transport network – its warehouses, bridges, hostels and customs sheds – he acquired political muscle as well. In 1648 the Duke of Savoy made him Baron Stockalper von Thurm, and his place at Europe's top tables was assured. From then on, he consorted with dukes, popes and princes.

In hindsight he looks very much like an early advocate of globalisation – a prototype of Davos man. The geological fluke that gave him rights over the mountain path handed him a multinational empire with interests in Antwerp, Lyon, Genoa, Paris, Rome, Sicily and Vienna, all run from a hill station in the Upper Rhone. He had a before-his-time faith in education and progress, too. After his six sons died young, leaving him with eight daughters, he became committed to women's education, and gave Brig an Ursuline convent with a college for girls. But his primary achievement was to embed the idea of trade in this wasteland of eagles and mountain deer.

Switzerland had until then been an impoverished land peopled by subsistence farmers working in extreme and difficult conditions. Suddenly, the country was a business hub. The bridleways became railway tracks and an economy based on transit was born. Indeed, Stockalper's affluence (it would have taken a peasant 370,000 years to earn the fortune he accumulated), along with the connections he forged, did much to establish Switzerland as a trusted go-between in mercantile as well as social matters. After taking custody of that princess, he extended the tradition of hospitality into something even more lucrative, but just as Swiss: financial stewardship.

Private banking is a gilded industry, but it too draws on the old idea of calm service – on what we might call '*hospizality*'. It is effectively an elite concierge service for money, and once Switzerland had made itself synonymous with looking after people's health, it was only natural that it should seek to look after their wealth, too.

Neutrality, stability, the ability to offer refuge in a storm . . . these all became tangible commodities. And when, at the end of the Napoleonic Wars in 1815, the Congress of Vienna enshrined Swiss neutrality as a geopolitical fact, it only added to its reputation for reliability. The senior European powers (Austria, Britain, France, Portugal, Prussia, Russia, Spain and Sweden) were united in wanting access to the Alpine passes, and since they wanted access to its bank vaults too, they were eager to grant it 'perpetual neutrality'. Switzerland could now present itself as a vulnerable island in a sea of absolute monarchies, and elevate that neutrality into its best and only defence.

Its allure in this area would later be badly damaged by its willingness to accept every sort of gold, no matter how ill-gotten. Those famous secrecy laws may have been codified originally (in 1934) to *protect* Jewish assets, but in the end they permitted banks to treat Nazi Germany as just another valued client (Adolf Hitler had a Swiss account) and to look the other way when it came to the stolen treasure that was filling its vaults. Despite some moves towards transparency, Switzerland remains, in the world's imagination, a secretive realm of nefarious wealth, safety deposit boxes and numbered accounts – a non-fiction Gringotts, hidden in deep caves beneath those beautiful snow-capped summits.

~

Switzerland's location gave it another remarkable feature that was advantageous on the international stage: three languages. And that multilingual character made it an intriguing microcosm of Europe as a whole, a Europe in miniature – indeed an experiment in what the continent might one day become: a multilingual federation based on co-operation rather than rivalry.

In a way it was the rivers that gifted Switzerland its languages. The mountain-fringed valleys of the Rhone, the Rhine and the Ticino were walled in by ridges that made it hard for the people within them to communicate with each other. This imposed isolation made them ideal incubators for different ways of communicating. The Alemanni, who pushed the Romans aside along the Rhine, planted German as they advanced into the hills.* The Burgundians who occupied the Rhone west of Lake Geneva carried French all the way up to the glacier. The Ticino filled up with Latin. From there, despite the fact that they were neighbours, each valley developed in a strikingly different direction. In Switzerland the cloud-piercing mountains kept those language zones firmly separate, but as part of a coherent Alpine union, which encouraged them to attempt something rather interesting and novel: co-existence.

In fact, there are four official languages (German, French, Italian and Romansch) and even a de facto fifth: English. Schools are obliged to teach at least two; the 200 members of the National Council and forty-six members of the Senate are encouraged to use their own, without any need for interpreters; four of the cantons are bilingual; and the newspaper in the watchmaking town of Biel/Bienne contains articles in both German and French.

This linguistic originality is especially noticeable in the mid-valley town of Sierre, where German gives way to French with much less fanfare than (for comparison) is visible on the Severn bridge

---

* It is thanks to this tribal migration, in the fourth and fifth centuries BCE, that Germany is known as *Allemagne* in French, *Alemania* in Spanish, and *Almania* in Arabic.

between England and Wales. But though there is no detectable dividing line, the change is both immediate and thorough. One minute the road is full of Volkswagens, the *Konditorei* offers *Gulaschsuppe* and *Aprikosenkuche*, the signs in front of the *Rathaus* point to the *Schule* . . . the *Polizei* . . . the *Kirche*; the next there are Renaults and Citroëns in the car park, the cafes smell of *croque monsieur* and *jus d'abricot*, and the *banque* is between the *tabac*, the *pharmacie* and the *boulangerie*.

This tactful arrangement has been hard won. Until Napoleon invaded the Rhone Valley and pressed it to become part of a French-speaking Swiss federation, Sierre was strictly German, and its children were punished if they dared use the language of Voltaire. Now they learn both as a matter of routine.

Embracing three languages had two far-reaching consequences. First, the effect it had on Switzerland itself, creating an unusual hybrid in the heart of a divided continent. Second, it offered a vision of what Europe as a whole could aspire to: a spirit of compromise, accommodating different cultures within one community. It is surprising, in this light, that it is not in fact part of the European Union, whose citizens, goods and services criss-cross it every day. Perhaps that is because it is already a mini-union on its own.

~

Switzerland's reputation for political stability and neutrality has endured. In time this made it an ideal home for countless international bodies: the United Nations, the Red Cross, the World Health Organization, the European Organization for Nuclear Research (CERN), the European Broadcasting Union, the Aga Khan Foundation, Médecins sans Frontières, the World Meteorological

Organization, the World Council of Churches and dozens of other such agencies built head offices in Geneva, and not only because institutions of this sort tended to be staffed by the kind of people who liked skiing at weekends.

One of the earliest of these associations was founded in 1863 by Henri Dunant, a scholarly businessman from Geneva who used to browse in the city library at the same time as Lenin. The horrors he saw in the aftermath of the Battle of Solferino,* which ended the Italian War for Independence in 1859, inspired him (if that's the right word) to create a humanitarian agency placed above national or military considerations. For its symbol he inverted the colour scheme of his country's flag and created the Red Cross, giving his brainchild a Swiss flavour it would never lose.

A year later he took part in the conference that led to the formation of the Geneva Convention, further cementing the idea of peace as a natural Swiss product. This won him the very first Nobel Peace Prize in 1901.

The drumming of feet on the banks of the Rhone created a market in tourism and trade, and then sponsored the financial and legal infrastructure required to support it. This led to huge advances in hospitality, engineering and medicine. Thanks to the route taken by the water gurgling out of the great glacier at the head of the valley, a rural backwater became a high-tech powerhouse.

---

* Solferino was the last battle in which the warring armies (in this case France and Austro-Hungary) were personally led by their monarchs. It is also the battle whose after-effects ripple through Joseph Roth's much admired novel *The Radetzky March*.

11

CAME, SAW, DIVIDED

The first springs of the Rhone, the Rhine, the Ticino and the Reuss mark the four corners of a high-level walking route named the *Vier Quellen* ('four sources'), which forms a ring around Andermatt. This is the central cone or pyramid (or what geographers call the 'igneous intrusion') from which the waters radiate outwards. Reaching these sources involves a bracing uphill march, though of course Switzerland provides photogenic trains and roads for the lazier sort of pilgrim. The Rhone is the furthest west, the Reuss and the Ticino guard the northern and southern wings, and the Rhine takes its first steps in a mountain lake to the east – Lake Toma.

When the sun shines, the lake has a Nordic tang, as if a chunk of Iceland has floated south to sparkle on this crowning Alpine massif. But on cloudy days the cliffs look wintry and the water is dark. If anything, it scarce deserves to be called a lake; it is a pond.

Nor is it anywhere near the beaten track. The trail starts by the Oberalp railway station – a stop on the Matterhorn–St Moritz line – where there is a restaurant, an information hut, a gift shop and a lighthouse, a jaunty replica of the one at the Hook of Holland where this water ends. The original is in a Rotterdam museum; this copy was hauled up the hairpins on the back of a lorry. But it does not feel out of place. The red light designed to flash at tankers on their way to Rotterdam works well for fearful walkers lost in these crags on a foggy night.

The hike up from the road then involves two tough hours along a rocky path steep and sheer enough to require walkers to use their hands as well as their feet. Yet on any given day, throughout the summer, people pick their way to the top. Some are grizzled veterans, clicking proficient walking poles, others carry dogs or infants in backpacks; some are soldiers, climbing for exercise; others (like us) are just tourists, gasping at the scenery.

All share a common purpose. Panting as they trudge, pausing here and there to have a drink and admire the (literally) breathtaking views over Heidiland, muttering greetings, they are engaged on the same quest: to reach the source of Europe's grandest river.

Flowing through six countries – Switzerland, Liechtenstein, Austria, Germany, France and the Netherlands – the Rhine was destined to be both a frontier and a highway, dividing and connecting Europe down the centuries. It attracted armies from all corners of the continent and the cities on both banks have changed hands many times. But above all, this stream of water became the embattled demarcation line between French and German dreams.

~

At first sight, the Rhine's beginning – known as the *Rheinquelle* – is not as poetic as the source of the Rhone. There is nothing to suggest the enormity of the civilisation-smashing wars that have raged over this little rivulet further down. It does not leap out of a glacier in a spray of great expectations. There must once have been ice on this shelf, for this is a glaciated cup, with high sides that would have

From the summit to the sea. A replica of the lighthouse at
Rotterdam stands guard at the top of the Rhine.

been scraped by a frozen spoon. But that was a long time ago. The Rhine now trickles out of its lake with little sign that it is anything out of the ordinary. The vista down the long valley is thrilling, of course, but most visitors, on reaching the crest, have eyes only for the *Rheinquelle*. A brass plaque alerts them to the fact that from here it is 1,320 kilometres to the sea – a rather un-Swiss misprint, since in fact it is 1,230 km. But either way (the guidebooks agree) this is the only place where Europe's grandest river can be crossed in a single step.

On the hump above the water, next to the monument, a *Rheinquelle* bench faces the view. And unlike the upper stretch of the Rhone – all scattered boulders and stunted trees – at first sight this is very much a *river* valley. Pine woods and meadows rise at angles above bluffs that bear the interlocking V shape of a landscape carved by running water, not ice. This is an illusion, however: this valley, too, was gouged by a tongue of primordial glacier. Indeed, the rift that curves away to the left in the far distance is the deepest north–south gap in the entire Alpine chain. An early cross-section of European history comes into view – this too was an ancient route from Italy to Lake Constance and the Roman Rhine.

Standing on this top shelf, it is humbling to contemplate the scale of what came from this small cold lake on this high windy mountain. And it is impossible not to imagine the route it would take from here: down to Liechtenstein, a relic of old Europe, before racing on to Germany. It is almost possible to hear the boom of the Rhine Falls, and to see the cathedral spires of Basel and Cologne, the castles, towers and vine-clad hills, the chimneys and towers of Düsseldorf and Duisburg.

As it leaves Basel, the river enters a sharply altered landscape. The tectonic collision that created the Alpine chain in the Cenozoic era (when the African and Eurasian plates ground into each other, forcing

Earth's crust to buckle upwards) also gave birth to two parallel lines of hills running northwards away from the central upheaval (a distant ripple of the same upheaval created England's South Downs too). Meanwhile, the gutter left behind became the valley of the Rhine.

The Vosges mountains stand on one side, the Black Forest on the other. Between them lies a broad rift valley some 40 miles wide. Here at last the river can slow down and curve through moderate terrain – in fact it ceases to be a river in the usual sense, but is more like a low wetland area, with shifting streams and gravel islands that have come and gone over the passing years.

Modern visitors can hardly fail to be struck by the thunder of traffic along the shores of this part of the Rhine. As on the Rhone and the Po, the river has historically been so unstable that human settlement has had to keep its distance. But people did need to be at least within range of the river for water, fish, energy and security, and the roads and railway lines followed by default, clinging to the easy ground. The result is that the once tranquil Rhineland is a juddering mass of vehicles. Trucks, cars and motorbikes roar along the autobahns, rusty goods trains rattle through stations in a scream of steel and fumes, and the water itself thrums with vessels: cruise ships, container ships, gas tankers and even rubbish barges grinding their way to distant incinerators.

It is no sort of bower, not the sort of place to lie down beside and read a poem. But somewhere in the clatter one can still feel the tremor of ancient feet.

Rome entered France from the south, pressing against the current, but in Germany it followed the Rhine downhill, a much easier journey. It arrived in the person of Julius Caesar, whose campaigns in Gaul were spreading Latin rule into the forested regions north of the Alps. Vexed by the tribal incursions from the German side of the Rhine, Caesar decided to confront the tribes with a display of Roman strength and drove them back to the river. Then, somewhere near Koblenz, he built a bridge.

It was the first of many. Rivers have always been critical strategic assets, and they are especially prized during conflicts, either as obstacles to be overcome or as natural lines of defence. Rome's engineers had no equals when it came to building bridges, and their constructions over Europe's dangerous and often impassable torrents played an important role in many of the empire's most spectacular campaigns.

The bridge at Koblenz was a good example. Plutarch's life of Julius Caesar (the leading source for Shakespeare's play) contains the most celebrated description of this episode.

> Of the barbarians that had passed the Rhine, there were 400,000 killed. The few who escaped repassed the river and were sheltered by a people of Germany called Sicambri. Caesar laid hold on this pretence against that people, but his true motive was an avidity of fame, to be the first Roman that ever crossed the Rhine in a hostile manner. In pursuance of his design, he threw a bridge over it . . . He drove great piles of wood into the bottom of the river above the bridge, both to resist the impression of such bodies and to break the torrent. By these means he exhibited a spectacle astonishing to thought, so immense a bridge finished within ten days. His army passed over it without opposition.

The rapid construction of that first crossing in 55 BCE is a story in itself – it took little more than a week. Plutarch's account goes on to suggest that though the forest tribes did sue for peace, it was evident that they were only preparing for another fight. Eighteen days after its construction, the bridge was removed, Caesar abandoned his plans and sailed off into the unknown, to Britannia – 'an island', as Plutarch put it, 'whose very existence was doubted'. It says much about the fury of the German warriors that they gave even Rome second thoughts about trying to subdue them.

Caesar had sufficient status to create a new calendar, with a month (July) named after himself. And the impression he made on his German adversaries was powerful enough for them to use his name for their own leader – Kaiser. But the most bracing thing in Plutarch's account is that grotesque death toll. The idea that as many as 400,000 'barbarians' were slaughtered as they fled back across the Rhine . . . it makes the first day of the Somme (in which 57,470 men died, the deadliest day in the history of British warfare) look almost within the bounds of sanity.

Yet only a generation later Rome came back for more. The Emperor Augustus, impatient with the incessant raids from the German side (which interfered with the peaceful flow of leather, timber, salt and other commodities), resolved to pacify the eastern shore and invited his twenty-three-year-old stepson, Nero Claudius Drusus, to teach the barbarians a lesson. In military terms Drusus was a novice, but he started impressively, gaining control of the Alps before turning his attention to Magna Germania. He built a chain of military posts along the Rhine – at Strasbourg, Speyer, Mainz, Bingen, Koblenz, Bonn and Nijmegen – and only then took his army across to the other side.

The invasion began in 12 BCE, when the Roman river fleet ferried four legions north to what is now the Netherlands before marching into Germania. And at first things went as planned. For the first time ever, a Roman army was able to spend an entire winter camped east

93

of the Rhine. Drusus's reward was to be given a triumphal procession through Rome and made consul, an honour he celebrated by spending the next two years establishing Roman governance over his conquered lands.

Three years later, however, in 9 BCE, Drusus fell from a horse and injured his leg gravely enough to lead to his death. He was granted a conqueror's funeral, with triumphal arch and all the trimmings but, deprived of his leadership, his army was ambushed in the Teutoburg Forest by Arminius, a German leader who had himself trained in Rome. The Romans were forced back to their garrisons on the Rhine for good. In the centuries that followed these became the limit or *limes* of Rome's empire.

At the southern end, on the road up from Basel, was Argenturatum (Strasbourg). Downriver, at the junction of the Rhine and the Moselle, stood Castellum ad Confluentes (the 'Castle on the Confluence', or Koblenz). To the north lay Colonia Agrippina (Cologne). And in the easternmost elbow of the Rhine stood Moguntiacum (Mainz). Its 12,000-seat theatre was the largest north of the Alps, and it also had the region's tallest tower, a commanding monument to Drusus himself that was squarely planted in the citadel overlooking the river he had crossed.

It was known as the Drususstein, and a chunk of it is still there, a rough brick column in the grounds of Mainz castle, like the stump of a giant tree or a trunkless leg of Ozymandias. And one does not have to murmur a bit of Shelley (*little beside remains . . .*) to see in its ruin both a comment on the frailty of empires in general and a reminder of Rome's enormous power. The conquest of Germania may have proved short-lived, and ended in failure, but the cities Drusus built along the Rhine are an extraordinary testimonial to the awesome might he once represented.

~

Following the failure of Drusus's campaign, Rome made few further attempts to establish itself on the other side of the Rhine. The river came to represent the edge or outer limit of Rome's European ambitions.*

Yet its actions here – attempting to conquer the forests east of the Rhine, and then building a string of fortified cities along it – more or less guaranteed that this part of Europe would become a contested frontier between rival civilisations for the next two thousand years. Imperial Rome might have believed itself to be erecting a border; in practice it was opening up a fault line between two fiercely different worlds. To the Roman-Gaulish realm on the western side of the Rhine, the river was a frontier; but for the tribal territories to the east it was central to their world – the water they fished in and drank from, their home, their road and the symbol of their hopes. This continued down the centuries: French geographers would dream of France as being a perfect hexagon, framed by fixed physical features, but German thinkers came to see the Rhine not as the border of their national identity, but at its very centre. 'The Rhine is Germany's river, not its boundary,' wrote the historian E. M. Arndt (author of a popular patriotic song, 'Des Deutschen Vaterland') in 1815. It had always been, and always would be, 'Papa Rhein'.

As we shall see, a fearful number of Europe's subsequent conflicts would be inspired by the tension created by this division: the Roman, Visigoth and Frankish conquests; the battles between the barons and bishops of the Middle Ages; the empire-building

---

* Though further south, and centuries later, Trajan and Marcus Aurelius would cross the Danube and succeed in turning 'Dacia' into 'Romania'.

advances of Charlemagne and Barbarossa; the religious upheavals of the Reformation; the *gloire*-seeking campaigns of Louis XIV and the Habsburgs; the Duke of Marlborough's Rhine march towards Blenheim; the back-and-forth pounding of Napoleon's armies; Bismarck's assaults in 1870; the incomprehensible horrors of the First World War; the incessant Franco-German tug of war over Alsace-Lorraine; Hitler's unopposed occupation of the left bank in 1936; the fiery bombardments of the Second World War. The ancient capitals of the Rhine – Strasbourg, Worms, Mainz and Cologne – were incinerated again and again (and again).

Much as we might wish to see the Rhineland as a peaceful region scarred by war, in truth it is a battlefield interrupted, every now and then, by a truce.

And at the centre of it all, the clash of visions over the Rhine was supported by the nature of the river itself. Its length, speed, size, course and character made it answer to both these contrasting aspirations: a longed-for boundary and the heart and soul of a nation on the rise.

12

THE CORRECTIONS

When rivers slow down on the inner curve of their bends they deposit whatever sediment they are carrying and in effect lay down new soil. This is how they begin to bend. They refresh the land even as they plough a path through it. But they also flood. And though the spills of recent years appear to have been made more extreme by man-made climate change, rivers have been crashing out of their banks since time began.

In 2019 the Ticino burst its banks as it hurried through the ancient city of Pavia, south of Milan, and in the summer downpours of 2024 there were four deaths when the river sluiced into the villages north of Lake Maggiore. Many of the farms and villages in the region have marker lines showing the height of the rising waters, and they tell a story of repeated floods throughout human history.

On the Swiss Rhone, meanwhile, the inevitable mood swing between summer and winter, combined with random fluctuations, means that the volume of water running out of the snow bowls has never been reliable. In cold spells it could slow to a trickle, and sometimes an excess of warmth could also dry it down to an echo of its usual self. But at other times it could be a torrent. And because the glaciated valley was broad and flat, when the water did swell, it did not follow a fixed route but drove towards Geneva like a runaway horse. From time to time, pressed by a surge from above, it would charge out of its bed, ruining harvests and knocking over the chalets near its banks.

The river's boisterous nature, and the many challenges this has imposed on the people who live beside it, has had serious consequences (some of them obvious, some of them less so) not just for this one valley but for the country as a whole, and has inspired many attempts to control it.

～

In medieval times the only defences were wooden crates weighed down with stones. In time these developed into temporary dams, with branches laid on top, but while these moderated the flow somewhat, they did little to address the larger problem. The valley was often a stagnant swamp – good for wildlife, uninhabitable for humans.

And every time humans slowed the river, they were also nudging it to drop more debris, thus raising the bed – in some cases by up to 3 feet.

Which meant more flooding.

The Rhone (known, in its German first flush, as the Rotten) overflowed in 1840, 1854 and 1860, drowning the entire 100-mile stretch from the glacier to Lake Geneva. In 1868 a heavy deluge led to an especially destructive surge. As in the debacle on the Dranse in 1818, the overflow swamped farms, roads, chapels, fields, villages . . . everything.

And just as the flood in the Val de Bagnes had broader repercussions, so this horror of 1868 called forth a new sort of a national response. Understanding that intervening on one part of the river would generate impacts elsewhere, three major Swiss cantons – Valais, Vaud and Geneva – agreed to take preventative action together to deepen the river bed, build stouter walls and clear the rocky impediments that grew and shifted with every surge.

This united front was still a novel approach for a nation that until 1848 remained a gathering of resolutely independent provinces. Napoleon had tried to merge them into a French-style republic, but in 1847 seven of the Catholic cantons pushed to secede from the predominantly Protestant union. It led to one of the most polite conflicts in the history of civil war: after a hundred deaths, the two sides shrewdly agreed to re-unite under a new constitution that made Switzerland a federal state.

Thus, while the deluge of 1868 was by most standards a dreadful calamity, it also went down in history as 'the flood that changed Switzerland'. The spirited waters of the Rhone taught the newly formed federation lasting lessons not just in the maintenance of its rivers, but in the management of itself.

Since it was soon clear that any changes to the Swiss Rhone would have knock-on effects on the river west of Geneva, this had wider implications too. The river's temper tantrums called not just for national action but for international collaboration, leading to the first tentative steps towards shared responsibility.

As part of the First Correction, in the latter part of the nineteenth century, a barrage was built below Lake Geneva, in France, to regulate the flow of the Rhone as it left Switzerland – one of the pioneering projects in the energy generation we encountered earlier. It took three decades to complete, but although it was (literally) groundbreaking, it did little to update the old formula of barricades and reinforced banks. It offered protection to the fields, farms and

The uncorrected Rhone cuts through Valais, near Sierre.

villages on the immediate flood plain, but it was only a short-term solution – it simply passed the surging water down to the next bend.

By the time of the First World War, it was obvious that this would no longer suffice. Urgent action was imperative, because the railway was coming, and in this landscape there was no option but to put it on the level ground beside the river. A more thought-out approach was needed.

The Second Correction, planned in the 1930s but delayed by the Second World War, aimed to straighten and deepen the entire channel, effectively re-engineering the Rhone into a canal that could speed water down to Lake Geneva as fast as possible. In fact, there was more to this than mere flood defence: the architects envisioned that this valley, with its exceptional sun record (the way the mountains funnel the wind creates a remarkable microclimate – Sierre likes to boast that it is sunnier than Algeria), had the potential to be a mini-California, bursting with peach, apple, cherry and plum orchards, not to mention vineyards. It was whole-hearted land reclamation, and it involved tucking the river away in a trench, hidden from view like a badly behaved relative banished to the basement.

Anyone driving up the valley from Lake Geneva today would barely know it is there. Only when the road crosses a bridge is the Rhone visible, and it plays little part in the social life of the region – one looks in vain for a riverfront cafe, a bar, a restaurant or a waterside balcony. There are no hiking trails, boat trips or water sports of any kind. Signs warn passers-by not even to *think* about taking a dip: the current is too strong and the water too cold.

Switzerland is not short of beauty spots, so perhaps it can afford to bury the Rhone out of sight in this way. It is also pragmatic, content to give gravel pits priority over scenery. But the result is that this part of the river is all business. Dredgers dig and trawl the bed for stones; power stations tap it for energy. The valley's most obvious feature (if we tear our eyes away from the superlative peaks and vineyards) are the pylons that march over the roads with enormous metallic strides.

In the short term, this reshaping of the Rhone was a terrific feat of Swiss engineering – equal to its gravity-defying expertise in tunnels, mountain railways, cable cars and bridges. The valley *did* become an agricultural powerhouse – and, thanks to the emergence of winter sports, a tourist nirvana too, with a string of chic resorts (Davos, Gstaad, Murren, St Moritz, Zermatt, etc.) that were among the most stylish in the world.

It was a stirring example of what technology could achieve. But once again it was only a temporary fix, because in the twenty-first century it began to seem that being an intensive farm and a transport route was of dubious merit. Indeed, the floods of 1995 and 2005 were caused by a new problem: intense summer rain, worsened by environmental disregard. As well as the damage being done by the burning of fossil fuels, by squeezing the river for power, the dams on the Rhone were eroding the riverbed in measurable ways, which was unhappy for wildlife too.

Another overhaul was called for – a billion-dollar scheme known as the Third Correction. Leaning hard into the hope that humans could live 'with' rather than 'alongside' the river, the Rhone was released from its straitjacket and allowed to recover some of its natural vitality. For a few miles above Sierre it was even allowed to flow through wild forest. And at the head of the valley no attempt was made to create a land fit for apricots. The area exposed by retreating ice was left as a marshy fen.

It was a risk. For half a century the Canton of Valais had been one of the sunniest and most prosperous places on Earth. None of the

The corrected Rhone near Martigny: a river used to run through it.

millionaires landing their jets in the ski resorts wanted it to change. But there was no avoiding the fact that this luxury came at a price. The correction of the Rhone had turned it into a dangerous thing: a non-diverse ecosystem.

So, as with the disaster in the Val de Bagnes, natural calamity proved to be two things at once: in one way it was ruinous, but it was also educational. The water that poured out of the glaciers above the Rhone and raced through Valais had never been peaceful: there was wisdom in the folk legends that saw it as a demon. And while the attempts to control it were in some respects successful, new problems kept emerging in their wake. The Rhone, like all rivers, had instincts of its own that could not be easily suppressed. It was necessary to seek new ways of co-habiting with the waters people both needed and feared.

13

THREE LAKES

When the rivers leave the mountains, they take a pause, as if collecting their thoughts before continuing their onward march to the sea. The result is three beguiling yet clearly different lakes: Geneva, Constance and Maggiore. They are notable holiday locations now, but they have also been the stages for war, commerce and intrigue. Each embodies the culture of the region in which it stands; each expresses a distinctive national flavour; each has played a palpable role in both the local history of the region and the larger story. It is possible to see Lake Geneva as a place of leisure and international discussions, Lake Constance as a religious-military parade ground, Lake Maggiore as a botanical whirl of semi-tropical greenery.

But they have plenty in common, too. These brimming pools of Alpine water, with their sun-warmed terraces and curtains of snow, brought something powerful and long-lasting into European life: the dream of escape. Their serenity and warmth helped lay down the very idea of 'getting away from it all'.

LAKE GENEVA

There are few more beguiling places on a sunny day than the north shore of Lake Geneva (or Lac Léman in French). There is a giddy amount to admire. Immaculate villages stud a sea of green vineyards that overlook uncannily blue water, with a frisson of snow in the sky. Rise early for a dip and the air is so exquisitely fresh, as morning sun streams over the summits, that it is hard to see where water ends and sky begins. The whole crescent from Geneva to Montreux resembles a lakefront mansion, with spotless deckchairs and sunshades arranged to face Mont Blanc.

It is easy to forget that the lake is really no more than a bulge in the Rhone, a holding tank for running meltwater. It is large enough to have its own miniature tidal pattern (a 2-inch rise and fall) but has been attracting affluent visitors for so long now that it carries few echoes of anything resembling untamed wilderness. On fine days the water is postcard calm, and the peaks are as placid as a stage set. It feels like a first-class spa: groomed, cosmopolitan and extremely expensive.

Switzerland is sometimes satirised as a cultural backwater. Orson Welles's slur in *The Third Man* that a thousand years of peace and democracy had brought forth nothing better than a cuckoo clock has long been a reliable put-down. It has no basis in fact – for one thing, the cuckoo clock is Bavarian, not Swiss – and some of the sneering can be put down to the desire to find *some* minor fault in a country that is otherwise so enviable. But it nevertheless feels like a significant truth. Europe has not often looked to Switzerland for the cultural expression of its hopes and fears.

It has made up for this by attracting legions of maestros from elsewhere. The home-grown Jean-Jacques Rousseau (born in Geneva) set *Julie, ou La nouvelle Héloïse* in a village above Montreux, and his heroine's first kiss, a heart-thudding break with decorum, made pulses race across the continent. But the writers and artists who followed him came from much further afield. Dumas, Dickens, George Eliot, Hugo, Nietzsche, Tennyson, Thackeray and many others all came this way.

Geneva was popular with British travellers because it felt like France but it wasn't Catholic. Edward Gibbon was a student in Lausanne in the 1750s, a good place from which to brood on Europe's past in *Decline and Fall of the Roman Empire*; Hans Christian Andersen wrote *The Ice Maiden* in Montreux; Dostoevsky thought up *The Gambler* in Vevey; Henry James booked Daisy Miller into the *Trois Couronnes* hotel (close to Anita Brookner's Booker-winning *Hotel du Lac*); Arnold Bennett wrote *The Card* in a lakeside villa. We cannot say that the setting cheered T. S. Eliot – he worked on *The Waste Land* in Lausanne in 1921–22. Tchaikovsky and Stravinsky both found inspiration on this hospitable shore.

Geneva was a place of political refuge too – like Britain, it was an island of relative freedom surrounded by police states, and it began to attract a new generation of intellectual refugees – a shelter for the likes of Marx, Engels, Kossuth and Mazzini. In 1915 a group of 'ornithologists' gathering in Zimmerwald turned out to be undercover socialists. Two lived in Switzerland already: Trotsky and Lenin had been scheming in Zurich and Geneva since the turn of the century. That is why Geneva was the setting for Joseph Conrad's *Under Western Eyes*, a tense tale of Russian émigrés in hiding, conniving and deceiving each other and themselves in the shining name of revolutionary ideals.

After the Russian Revolution it was the enemies of communism who came to the lake to lick their wounds and hide their money in a hospitable and neutral sanctuary. And in the Second World War Switzerland in general became a place to which people escaped, whether it was the Von Trapps fleeing Nazi Salzburg or British POWs breaking out of Colditz.

In more peaceful times, a throng of other stars – Graham Greene, Noël Coward, James Mason, Freddie Mercury, Vladimir Nabokov, Peter Ustinov – retired contentedly to these parts (if only for tax reasons). The Argentine Jorge Luis Borges once called Geneva 'the most propitious for happiness' of all the cities he knew. John Russell, in his 1950 travelogue, referred to these well-heeled expatriates as 'the Emperor Moths of Western civilisation', and saw them as enjoying 'the eternal July of their leisured existence'. The most feted of them all, perhaps, was Charlie Chaplin. Born to an impoverished music-hall family in south London, he was hounded out of Hollywood by the conspiracy theorists of McCarthyite America (who read his support for Stalin in the Second World War as a sign of Soviet leanings, even though Roosevelt and Churchill had inclined the same way) and fled to a superb mansion on the hill above Vevey, with gracious rooms, huge lawns and an unbelievable view across the lake to the Dents du Midi.

It was (and still is, as a museum) a spectacular property, and Chaplin lived there contentedly with his wife – daughter of the playwright Eugene O'Neill – and children for his last years. Thanks to his eighty-film *oeuvre* and the radiant popularity of his put-upon Tramp character, he was one of the world's richest men (when he died he was worth an estimated $100 million). In his Vevey retreat he led the life of a lordly monarch in exile, writing his memoirs and socialising with other members of the A-list: Zhou Enlai, Einstein, Gandhi, Khrushchev, Nehru and many others.

Lake Geneva, where the Rhone flows into the sky.

But this Switzerland had something more than just superb scenery. As a calming model of *hospiz-ality* it exerted allegorical pressure too. Hans Castorp, the hero of Thomas Mann's *The Magic Mountain*, becomes so infected by the drowsy allure of the 'rest cure' in Davos that though he was intending to pay only a brief visit he ends up staying for seven years – long enough to cut himself adrift from life on the plains below.

Chaplin felt much the same way. 'I cannot tell you how happy I am,' he wrote to his brother. 'It is like a ton weight lifted off my shoulders: no fears, no hate, no blackmail, no politics.'

It wasn't just artists; business came here too. It is not known what drove (or led) Henri Nestlé to move from Frankfurt, where he was born in 1924, to Vevey, a Lake Geneva beauty spot. It was hardly convenient. But Nestlé wouldn't have been the first CEO to locate the corporate headquarters where he wanted to live, and Frankfurt's many qualities do not include a lake view of mountains. When he sold his company in 1875, he retired along the lake to Montreux – a sign, perhaps, that it really was the idyllic locale rather than mercantile considerations that brought him here in the first place.

Nestlé began as a chemist, authorised to blend and sell medicines, but soon branched out into lighting oil, fertiliser and alcoholic drinks. Then he devised a way to make milk powder. Industrialisation was creating huge cities with limited access to fresh milk, and Nestlé's powdered range was an immediate bestseller. It gave birth to a range of products that remain ubiquitous today – Nesquik, Nescafé, Nespresso and more.

There is a similar story on the other side of the lake, in France. In 1789, the year of the Bastille, the French Marquis de Lessert, walking above the spa town of Évian-les-Bains, quenched his thirst with water from a spring on land owned by a Monsieur Cachat. It was so refreshing that he decided to bottle it. The French Revolution put a stop to all that, but in 1826 a Monsieur Cachat started supplying Évian water in earthenware pots. In 1859 the Société anonyme des eaux minérales de Cachat was incorporated, and the mineral-rich water of the Rhone Valley found itself driving a major new industry.

Évian water spends fifteen years seeping through the rocks, washing itself in magnesium, calcium and sodium, before arriving at the bottling plant near the lake. Offering itself as a taste of the French Alps, the business (now owned by Danone) sells 8 million bottles a day, while its scientists try to devise ways to coat them in carbon-neutral plastic.

Hospitality, winter sports, chocolate, coffee . . . the waters of the Rhone, even when resting in Lake Geneva, have nourished many Swiss and French arts.

LAKE CONSTANCE

There is only one language spoken in the region around Lake Constance:* German. But it is fringed by three countries – Switzerland, Germany and Austria – and the tourist board claims that it actually has four parents, and calls itself the *Vierlandregion* ('four-man's-land'), on the grounds that Liechtenstein is only half an hour upriver. The ups and downs of the interested duchies (Baden, Bavaria, Berne, Prague, Württemberg) means that it has had as many as five different time zones. The exact boundaries have never been agreed.

---

* Known as Bodensee in German, it is the third largest freshwater lake in Europe.

In the past this led to confusion over timetables, and some of the clocks in some of the harbours bore more than one face, much as today's trading rooms display the time in the stock exchanges of New York, London and Tokyo. Trains failed to synchronise with boats. Only in 1904 did Lake Constance sign up to Central European Time, the standard embraced by Western Europe from Spain to Poland, and by the swathe of West Africa (Algeria, Morocco, Tunisia, Nigeria, Ghana) immediately to its south.

In the south-east corner stands the Austrian town of Bregenz, administrative capital of the Vorarlberg. It operates cruise ships on the lake and is home to a spectacular opera festival in the summer, with a 7,000-seat auditorium facing a stage that juts into the water in front of a glorious backdrop. The old Roman camp sits at the eastern tip, where the Rhine arrives; on the west stands the medieval capital of Konstanz. These are linked by the same distended wash of the river, a broad hollow scooped out by the Rhine Glacier, which swept out of the Alps 10,000 years ago. The western finger of the lake is actually called the Seerhein (Rhine Lake), and points due west like a cannon aiming at Strasbourg.

This is the German Riviera, a south-facing hideaway with lidos, beaches, fishing grounds, bird-spotting huts, bike trails and gardens. It is sprinkled with both ancient echoes and modern hotels. But it is also a lesson in the raw power of geology, because the Rhine did not always come this way. In prehistoric times it pushed on north to the Danube, only to be diverted by a geological commotion in the last ice age, something like 15,000 years in the past.

In this age of frictionless travel, it does not feel like a border – but it is one, because, as on the Rhine, this is where the Romans stopped. They maintained a fleet on the lake to keep the tribes of Germania confined to the northern shore and planted vestiges of themselves along the south.*

That northern fringe is thus thoroughly German – a string of Bavarian towns. The tourist highlight is Lindau, with its quaint island, harbour, lighthouse, churches and parks. Friedrichshafen is the industrial centre, and Konstanz the cathedral and university city. This last owed its eminence to the fact that for centuries it was the only bridge over the Rhine in this region. In 1414 it was host to the Catholic Council that spent four years trying to repair the schism between its Byzantine and western halves, and the first tremors of the Reformation were felt there, too: the Prague preacher Jan Hus was denounced as a heretic in Konstanz and condemned to death by fire in the main town square.†

By German standards Konstanz is relatively unspoiled – thanks to its southerly position, it was spared the worst raids of the Second World War, and the historic old town is, by German standards, unusually well preserved. But some of this was down to canny local politicians, who declined to issue blackout instructions when the air-raid sirens sounded, thinking to fool the Allied navigators into thinking that the brightly lit target below their open bomb-bays must surely be in Switzerland.

It almost is: it is a splinter of antique Germany surrounded by Swiss terrain. That gives it a modern role as a cut-price outlet for Swiss shoppers, who can raid its malls using the dramatic power of the Swiss franc. The currency advantage makes it so tempting that

---

* Indeed the lake owes its name – Constantia – to the Roman emperors who built their fortified encampments along its shores.

† In 1415 – the year of Agincourt. 'You may kill a weak goose,' he supposedly said, refusing to recant, 'but more powerful birds, falcons and eagles, will follow.' His ashes were disposed of in the Rhine.

many retailers have clustered here, and its notoriety as a duty-free emporium has been boosted by the German government's policy of offering VAT refunds to non-EU citizens. A scheme designed to tempt Chinese visitors into German malls has ended up subsidising people who live within walking distance of the till.

This catches the super-ordinary mood that surrounds the border between Germany and Switzerland in today's world – it is a well-favoured locale for shopping trips and retirement homes. But it has not always been so calm. Indeed, the most famous child of these parts was a warrior: Count Ferdinand von Zeppelin. After attending military school in faraway Ludwigsburg, on the Rhine, he joined the army as an engineer, but a taste for adventure took him across the Atlantic to North America, where he joined the Union army as a Civil War observer. One day a German-born balloonist named John Steiner took him into the skies over Minnesota, giving him a first taste of the idea that would shape his life.

After serving in the Franco-Prussian War he conceived a new sort of aircraft: a propeller-driven gas dirigible that could carry passengers great distances, but also be used as a bombing platform. His Zeppelin Aircraft Company was based in Friedrichshafen, and on 2 July 1900 his first Zeppelin went on a wobbly twenty-minute flight above the lake – over a sea of upturned faces. It was not a great success, but he was not deterred, and by the time the First World War broke out his airships had flown more than 30,000 passengers across Germany in perfect safety.

At that point, however, it was converted into a long-distance weapon, and before long was dropping high explosives on France, Belgium and even Great Britain. In 1915 the cigar-shaped bombers were appearing over Norfolk, the Humber, London and Dover – in all there were more than fifty Zeppelin raids on Britain, the largest

Lake Constance: a playground of five time zones.

of which, in 1918, was the last: three of the so-called 'Giants' joined a fleet of conventional two-winged bombers in a mass attack on London. There was little in the way of defensive artillery at that time – it had never been needed before. So, although one Zeppelin was brought down near Bristol, enough bombs were dropped to inflict grave damage, kill forty-nine people, and frighten the local population.

The Zeppelin factory was closed down by the Treaty of Versailles in 1919 – a setback the founder did not live to see; he died in 1917. But it was revived in the 1920s and became famous for the *Hindenburg*, an emblem of German renewal under Hitler until the disaster of 1937, when it caught fire and crashed in New Jersey, producing newsreel images so dramatic that public confidence in the airships was completely shredded.

Alongside the souvenirs of war, however, stand relics of peace. The outstanding feature on the Swiss shore is the ancient abbey of St Gallen. This part of the Rhine was one of Europe's earliest main roads, from Zurich to Italy. It was here, in the seventh century, that a wandering Irish monk named Gallus (later Gall) fell ill and retreated to a hermitage in the woods above the lake.* According to legend he was more than an energetic teacher; he had a miraculous ability to deliver people from demons and even made friends with a bear. This made him a natural choice as Bishop of Konstanz when the position became free.

The solitary life seemed to suit him: he lived until he was ninety-five. An abbey was built in his memory two centuries later, which in time developed an extraordinary baroque library of 160,000 old

---

* In some accounts he was not part of the Irish monastic tradition at all, but from Alsace. But he was certainly travelling up the Rhine in Bregenz when he stopped.

volumes and rare manuscripts. In honour of its Celtic roots, it has a fresco of Durham's Venerable Bede and a larger collection of Irish books than anywhere in Dublin, including rare medieval gospels. And thanks to the river running through the lake, it also has a manuscript of the *Nibelungenlied*, the grand poem of the Rhine in which Wagner found his gods and heroes (which we'll come to later).

When the river winds out of the lake it flows through the chocolate-box town of Stein-am-Rhein, a beguiling mass of medieval beams and gables, unusual in that it lies on the north – or German – bank. A trading post in Renaissance times, the presiding saint of its monastery church gives the town its crest – St George and the dragon – and since that was a legend brought home by crusading knights, it is a sign that this was also the way from the Holy Roman Empire to Jerusalem.

From there it is a short run to Schaffhausen, also on the north shore, home to a railway junction and a hydroelectric energy station, and a watchmaking capital. Timekeeping, of course, is another traditional cog in Switzerland's precision-tooled economic identity, but in this case it has ironic origins in Jean Calvin's puritanical suppression of luxury in 1541. Geneva's newly redundant jewellers and goldsmiths turned to watchmaking instead and founded a great national industry. It produced the first wristwatch, the first waterproof watch, the first quartz watch and many other breakthroughs – none of them cheap. And when classic timepieces looked set to be made obsolete by digital technology, Switzerland contrived to retain its hold on both the prestige end of the market (Rolex, TAG Heuer, Patek Philippe) and (through the Swatch) the mass market too.

Schaffhausen remains a town of time-honoured abbeys and guildhalls, some of which carry darker echoes. In 1349 its Jewish population was roasted alive after being accused of spreading the plague. Twenty years later the Jews returned, but in 1401 they were attacked again. This time, accused of 'instigating' murder and spilling Christian blood, they were first tortured, then publicly burned. Thereafter they were required to wear coloured caps as a badge of their identity (as in Venice, in the original Jewish ghetto). The beautiful town square also has a sobering plaque to the Jews who were murdered here in medieval times. This was part of the larger pogrom blighting all Europe at this time. Jews were being massacred in the Rhineland, too, in the crusading fury that engulfed Christendom, and were expelled altogether from England, France, Spain and Portugal.

Many centuries later Schaffhausen became the only Swiss town to be bombed by the Allies, on April Fool's Day, 1944; a hundred people died in the night raid. Fortunately, given that it was one of the most picturesque towns in this very picturesque region, the medieval centre, with its fairytale gables and crooked beams, was undamaged. The US Air Force apologised and insisted that it was a mistake, but when it happened again the following February (killing sixteen) suspicions grew that it might have had something to do with the fact that the munitions factory was rumoured to be doing brisk business with Nazi Germany.

After Schaffhausen, the river gathers pace until it splashes over the fabled Rhine Falls – the inspiration for a thousand melodramatic paintings. This is where hard Alpine rock meets lowland gravel, and the churning water has chopped out a cliff that now generates a crash of white water. With a 75-foot drop, it does not compare to the cataracts of Niagara or Victoria Falls. But by European standards it is epic. The river is 500 feet wide at this point, and the whole flood hurls itself over the ledge in a bravura plunge that has drawn artists this way for centuries. Turner's watercolour is only the most famous of a thousand such images.

A jutting two-fingered rock – the Rheinfallfelsen – stands in the centre of the stream, leaning into the flow as if fighting to hold back the surge. A remnant of the original ridge, it is now a convenient viewing platform. Visitors can stand in the heart of the thundering river, sheets of water flying above and around them. When Mary Shelley visited, she reported: 'The spray fell thickly on us . . . looking up, we saw wave, rock, and cloud, and the clear heavens moved through its glittering, ever-moving veil.'

On the north bank stands a seventeenth-century mill, built to power an ironworks, but the Rhine Falls have not been tapped for hydroelectric energy in modern times. This spectacle of foam and sky, it turns out, is worth more in tourist francs than anything a power station could bring.

LAKE MAGGIORE

Lake Maggiore differs from the other two lakes in that it is narrower and fringed by steeper hills. It wriggles out of the mountains like a serpent, making Lake Geneva and Lake Constance look like placid reservoirs. It is therefore faster – it takes a dozen or so years for a single bucket of water to pass through Lake Geneva, but only four to run through Lake Maggiore – and deeper, with a maximum depth of 1,220 feet, against 1,017 for Lake Geneva and 823 for Lake Constance. It was created by an enormous glacier, a thousand feet thick, that pressed out of the mountains like a battering ram. The moraine it pushed up at its tip became the dam behind which the lake formed. If it weren't for the blue sky and brilliant flowers it would feel like a Scottish loch. But that colour scheme changes everything. Even the northern end, where the Ticino comes out of the mountains, feels utterly Mediterranean, all terraced gardens, sprays of colour, hot-country foliage and lemon groves. Not for nothing are Locarno and Ascona known as the Swiss Riviera. The shore and its islands are bursting with luxurious villas and balmy botanical grounds.

The sheer allure of this terrain had obvious repercussions. Patricia Highsmith, the poet of American crime writing, lived just off the Ticino near Ascona (in a rather gloomy house). Lake Maggiore was popular with politicians too. Winston Churchill and Helmut Kohl were regular visitors, and Queen Victoria enjoyed a famous stay in the Villa Clara at Bavona – there's still a watercolour of the view in the Royal Collection.

Henry James spoke for many when he wrote, on seeing Lake Maggiore: 'One can't describe the beauty.' And when James Joyce, hard at work on *Ulysses*, heard about the murals showing scenes from Homer's *Odyssey* on the walls of a fashionable summer palace on the island of Brissago, in the lake near Ascona, he had to go there himself to take a look.

From 1900 Ascona was home to a flamboyant artists' colony: Monte Verità. Based in a former seminary, this was a spiritual play-pen set up by exiled intellectuals including Mikhail Bakunin, the Russian revolutionary, and Rudolf Steiner, the avant-garde educationalist. In time it turned into an anarcho-vegetarian utopia for naturists, then a creative retreat for the likes of Herman Hesse (*Steppenwolf*), Erich Maria Remarque (*All Quiet on the Western Front*), Paul Klee, Isadora Duncan and many other musicians and painters. Summer on the lake had never been so giddy. Switzerland was hardly known for bohemian excess, so this was radical stuff. And traces live on: the bookshop beside the lake has roomy walls of high-minded literature in four languages – no mean cultural achievement.

Lake Maggiore, where Switzerland flows into
Italy. 'One can't describe the beauty'.

Swissness triumphed in the end: the once risqué home of nude dance therapy is now a reassuringly expensive hotel – a little touch of Bauhaus on the lake. And that perhaps helps us see, in retrospect, that the carefree lifestyle of Monte Verità may have been merely a child of the exchange rate. In the interwar period a British pound would buy 20 Swiss francs – making it a refreshingly cheap place to enjoy a free and uninhibited existence – much superior to the philistine grind back home.

Today, with the Swiss franc so mighty, the north end of Lake Maggiore has become all but unaffordable. But its role as one of Europe's premier pleasure grounds has led to its reincarnation as a sun-soaked campus for wellness getaways and spa treatments. Those daring bohemian holiday camps have become the modern equivalent: weight-loss clinics. In this they extend (perhaps unwittingly) the fine old Swiss tradition of hospitality. One of Europe's strongest motive forces – the sybaritic pursuit of pleasure – found a natural home on these Italian lakes.

Not that there aren't national differences. Ask the concierges in the Swiss hotels what time the ferries run, and they are likely to spread their hands and say that in the old days they could have told you, but these days, now that the ferries are run out of Italy . . . (shrug) . . . who knows?

On the surface, Maggiore seems untroubled – an idyllic paradise it is hard to imagine ever being blighted by anything unseemly or rough. At the foot of the mountains, beneath the Titian-blue Italian sky, north meets south in a languid blur that seems entirely and literally cloudless.

But there *have* been noisy episodes. In 1873 a dynamite factory opened in Ascona, licensed to manufacture Alfred Nobel's explosive new invention for the engineers up at the Saint-Gotthard.

The project needed a local supplier, dynamite being too unstable to tolerate long-distance travel. But the following year a series of shattering accidents blew out the windows of Ascona and the municipal authority expelled the business. First it went to the island of Brissago, in the middle of the lake, but eventually settled further south. There it produced half a million kilograms of nitroglycerine, more than enough to blast a hole in the Alps.

The roars may have been hideous, but on this occasion the cause was peaceful – a railway tunnel designed to enhance free exchange. And it worked. The primal logic of its route through the passages carved out by the rivers went on to help Lake Maggiore play a famous part in the peace process after the First World War, when Europe's leading powers met in Locarno to iron out the tensions that Versailles had not resolved. It was an ideal setting for such a conference. Germany was still reeling from its hyper-inflationary death spiral of 1923 (the price of bread rose from 250 marks in January to 200 *billion* by December), while France could not bring itself to forget the 1916 Battle of Verdun in which 160,000 of its young soldiers had fallen.

Where better to seek reconciliation than in a Swiss hotel, on an Italian lake, negotiating the future in three or four languages, in mild autumn sunshine and in the gentle embrace of Switzerland's reputation for calm neutrality? The 'Spirit of Locarno' became a tangible force in world affairs, creating the optimistic mood that led, in time, to the formation of the League of Nations. It even earned Sir Austen Chamberlain, the British Foreign Secretary who led the negotiations (and was Neville Chamberlain's half-brother), the Nobel Peace Prize of 1925.

All of this made Lake Maggiore synonymous with the idea of peace-making. So it was no surprise when, in 1935, the most important Italian town on the lake, Stresa, hosted a follow-up meeting. This had been a popular resort since Victorian times, with a garden island created by the Borromeo family, the seventeenth-century power in these parts. And Stresa's Grand Hotel des Îles Borromées was where Hemingway had staged the climax of his 1929 novel *A Farewell to Arms*. He had sneaked his lovers, warned that they faced arrest, into a boat and made them row all night up the lake to Brissago, there to gorge themselves on freedom and Swiss jam. It was an adventure that seemed to capture precisely what Lake Maggiore represented: the thrill of freedom.

In the long run, neither of these treaties held. And as if to show that nowhere could think itself immune, a few years later war did come to Lake Maggiore. There were no pitched battles, but in September 1943, after the armistice ended the alliance between Italy and Germany, a Nazi Panzer division arrived and the atmosphere changed. This remote area had not yet surrendered its Jews, and one lakefront hotel, the Meina, was a well-known shelter. Run by a couple from Constantinople, it had been housing Jewish refugees from Greece for years. But on 15 September the hotel was surrounded by the SS and its guests were seized, led away, shot in the head, and tossed into the lake – 'atrociously sacrificed', as one memorial plaque put it, 'to an unspeakable myth'.

That myth claimed fifty-four victims. The oldest of the bodies in the lake was seventy-six; three of the others were children.

In the spring of 1945 a secret meeting took place in these same surroundings: in a tense parody of the spirit of Locarno, Allied leaders met German generals at a villa in Swiss Ascona with a view to hastening the end of the war in Italy. It was named Operation Sunrise and was very hush-hush – with German-occupied Italy only a few miles down the lake there were obvious dangers. One of its leading figures was Allen W. Dulles, director of US Intelligence in Switzerland and later head of the CIA (an airport was named after him in Washington).

It was a success. On 2 May the shooting stopped. And the lakeside villa where the secret handshakes took place is these days the restaurant of the upmarket Eden Roc hotel.

This beautiful lake is the setting of mysterious episodes too. In 2009, for example, an avid crowd gathered to watch a vintage car being hoisted out of the water near that very hotel. It was a 1925 Type-22 Bugatti roadster, and it had been buried at the bottom of the lake since 1936. How it came to be there no one knew – some said the owner was a wealthy architect, others that he was a playboy gambler. Whoever it was, it seems that the car was deliberately scuttled in order to avoid the heavy Swiss import taxes. In time the chain securing it in place rotted away, and the Bugatti sank into the depths. Local divers found it, mounted an expedition to drag it out, and watched in amazement as it went on to fetch $364,700 at a Bonhams' auction (thanks to a classic-car museum in California).

It was a proud moment for the Belgian motor parts company, Engelbert, which had made the wheels – after all this time, the 'Bugatti in the Lake' still had air in its tyres.

Even more mystery surrounded the sinking of a pleasure boat in 2023. Four people drowned in Maggiore's southernmost finger when a storm from nowhere capsized what the papers called a 'birthday cruise'. When it turned out that two of the casualties were in Italian Intelligence, one was a retired Mossad agent and the other – the wife of the boat captain – was a 'Russian woman', tongues soon began to wag. And then it transpired that *all* twenty-three passengers were linked to the Italian or Israeli security services. Many of them made themselves scarce in the aftermath, abandoning their hire cars as they hurried to the airport, where business jets were waiting.

The *Corriere della Sera* wasn't the only paper to find this suspicious. The accident resembled something out of a spy novel. Where had all these interesting passengers been staying? No one knew.

~

The fact that pleasure-seekers have always loved this Alpine scenery has had substantial consequences in the form of its tourist industry – between them, the three lakes attract 30 million visitors a year (measured in overnight stays). But there is more to that than mere economic advantage. The way that the rivers expand and unwind into these welcoming blue lagoons encourages humans to imitate them by doing exactly the same thing themselves. Just as a river can be a metaphor for our own lives – representing the circuitous journey from youth to old age – so these lakes reassure us that life itself can slow down; that we too can get off the treadmill of our ordinary lives, exhale, tilt our faces up to the sunshine and enjoy a well-earned pause, lull or refuge from the daily rush.

The very idea of *vacation* implies a desire to empty ourselves. So the long history of sightseeing in these parts, that proud culture of peace and leisure – early Grand Tourists, bohemian summer camps, lake cruises, panoramic trains, lakeside spas and clinics, palace hotels and eccentric adventures – was in part inspired by the way the rivers relax into these glaciated bowls, to gather themselves before marching on again.

14

LOCKS
AND QUAYS

South of Lake Maggiore the Ticino (now the Fiume Ticino) enters a hot, low plain that has little in common with the Alpine precipices at the lake's northern end. We might think that it would become less pushy as it enters the pancake-flat world of Lombardy and the Po. But such is the pressure from above that the Ticino is a skittish character, darting and scrambling through the woods and rattling the stony banks as it bustles down to Milan and then on to its confluence with the Po.

Thanks to its unreliability, the river has managed to retain a wild and unspoiled character – much of it is protected, a national haven for deer and otters. Few buildings cling to its banks, and the townships have wisely kept their distance. In 1884 a barrage named the Panperduto ('a jewel of industrial hydraulics', according to the leaflets) was built across it to calm down the flow: the name means 'lost bread', and refers to the cargoes that used to be engulfed by its bounding currents. The river now runs through green woodland with fast water babbling on the pebble bed. Kingfishers twist their heads on low branches, and kestrels hover.*

In many ways it would have been logical for a great city to have clustered on the stream, with bridges allowing it to occupy both sides; but the historic capital of the region, Milan, preferred the firmer ground 20 miles to the east, midway between the Ticino and the Adda. Its original Latin name was *media plano* – middle of the plain – and it has been a political, religious, cultural and commercial centre. German kings, Swiss militias, Austrian emperors and French generals have all vied for control of this fertile territory, not least to secure the routes over the Alps.

But there was a problem. Although the water table was conveniently close to the surface in this low-lying area (Milan is only 400 feet above sea level and it has 433 wells), it urgently needed better access to water for transport. That is why it was forced to become, unusually for a place so far inland, a city of canals. It could never have become Italy's second city (after Rome) without this fan of new waterways. It may not have been actually *on* the river; but the Ticino was very much its parent.

The first of the canals, built in 1179, connected Milan to the Ticino at Tornavento, 20 miles west of the city, close to where Malpensa Airport now thunders. Originally a drainage ditch known as the Ticinello (after the river), it was later renamed the Naviglio Grande (in order to echo Venice's Grand Canal). It funnelled water out of the Ticino to create a highway running straight to the basin of quays at the Porta Ticinese.

---

* Alas, it is not entirely the haven of peace this implies, being close to Milan's Malpensa Airport, which handles more than 40 million passengers a year and 750,000 tonnes of freight. It has wonderful views of the Alps, though.

One of its first duties was to ferry the translucent marble (white, pink and grey) from the quarries of Condiglia, in Piedmont, needed to make Milan's extravagant Gothic duomo, the third-largest church in the world after Rome and Seville, and an architectural wonder.* The Piedmont marble that went into the wedding-cake monastery at Certosa, between Milan and Pavia, also came down the Ticino from Lake Maggiore.

These days the Naviglio Grande is purely recreational, a buzzy home for rowing clubs and pleasure boats. The restored docks have become the city's trendiest area: a colourful bustle of bars and cafes, where passers-by can sip their Aperol and marvel at the crystal clarity of the water, shivering with fish and the mazy reflections of churches. But during the Middle Ages this was the main spoke of Milan's commercial life, linking it to Pavia, Piacenza, Cremona, Padua, Mantua and the world. The docks and quays were the creaking heart of the city. Until its waters were paved over in the road-building craze of the 1960s, Milan was – to the surprise of many visitors – one of Italy's major ports.

The Naviglio Grande to the Ticino was an important first step. But in 1498 the Duke of Milan, Ludovico il Moro, saw that something more ambitious was needed, and appointed a state engineer to oversee a new canal system. The man he chose was . . . Leonardo da Vinci.

He was not a child of Milan; as his name implies, he was born in Vinci, near Florence. But he had spent two lengthy periods in the city, at the end of the fifteenth century and the beginning of the sixteenth, and made some of his most famous works here – not

---

* Work began in 1386 (a stone commemorates the moment) and continued until 1965 – a remarkable amount of time, we might think.

least *The Last Supper*. It was in Milan, too, that he had revealed himself to be a polymath, absurdly gifted in science, engineering, anatomy, poetry and a great deal else. More to the point, having seen the Arno bursting its banks as a child, he had a long-standing interest in hydrology. He had once proposed to weaken Florence's rival city, Pisa, by diverting the Arno away from it, and went so far as to urge Venice to create a unique defensive system based on self-flooding. His sketches contain numerous brainwaves for irrigation and drainage. In a book he called *Di mondo de acqua* ('Of the World and Its Waters') he transcribed his thoughts in a series of notes and sketches that still feel revelatory. His presiding insight is perhaps best summarised by its opening lines: 'Water,' he writes, 'is the driver of nature . . . the vital humour of the terrestrial machine.' It 'wears down the high summits of the mountains' and circles the world.

At the time, Leonardo had just finished painting *The Last Supper* on the refectory wall of Santa Maria delle Grazie monastery, but he was better known as an inventor than as an artist. His notebooks bulged with bold drawings: flying machines, gun turrets, mills, submarines, perpetual-motion toys, paddle boats and bridges. He imagined a 'perfect city', with a network of pipes that resembled human veins, and performed the same duty, sweeping nutrients – goods, people and ideas – around the city while flushing its gutters and disposing of its waste.*

Since Milan did not have a river, he would have to build one.

He began by organising the existing waterways into a cohesive system, and added a new one – the Martesana – to the east. The

---

* The outlines for these and other innovations are in a Leonardo Museum in Milan's Galleria – a display of science and engineering devoted to this maestro of design.

Leonardo's lock: Milan's canals linked it to the rivers and the sea.

ground was uneven, though, and the old medieval lock gates – heavy doors that had to be raised and lowered like guillotines – were cumbersome. Leonardo had an idea.

His solution seems simple now. It involved creating a gated pool that could be rapidly filled and emptied, allowing boats to climb or descend a step made of water. The hinged gates were angled upstream so that the force of the stream could seal them tight, with small hatches at the base, which, when opened, would let the higher basin flow more easily. Suddenly, groaning teams of men and mules were a thing of the past.

It is not known whether Leonardo's design was widely deployed in his lifetime. But his layout has been copied by canal engineers ever since. Not many of the bargemen who climb the famous 'staircase' on the Kennet and Avon at Caen Hill, in Wiltshire, tip their boathooks to Leonardo. But they should. So should the tanker skippers in the Panama Canal, whose giant lock gates obey the same Renaissance principles.

That is why the fragment of the Martesana off the ring road in Milan's business district is so evocative. It lies a short walk south of the railway station, a Mussolini-era showpiece built in 1931 to replace the 1864 terminus for trains coming through the Simplon. Two timber gates face each other along a stone chute, below an empty holding pool filled only with the chatter from a nearby *trattoria*. Those wooden barriers are battered relics, remnants of a larger apparatus. The empty basin is no longer than a cricket pitch, and no longer serves any aquatic purpose.

This was the artery that connected Milan to the Adda (which ran along its eastern flank before also joining the Po), allowing it to trade with the dairy farms of Bergamo and the silk looms (this was the beginning of Milan as a fashion capital) and fisheries of Lake Como.

Hidden below the Dubai-style spire of the UniCredit building (Italy's tallest office block), the empty culvert is thus more than a monument to a famous maestro. Milan could not have prospered without its waters. It would certainly not be what it is today: the capital of northern Italy.

15

RIVER CROSSINGS

The meeting place of the Ticino and the Po is a peaceful spot. An iron bridge originally built for the railway now carries cars across the rivers, over islands and shoals, with a marina on the northern bank, a restaurant and almost no other sign of human life. Birds rustle in the long grass.

The scale of the waterway might astound a first-time visitor – it is half a mile across, as broad as an estuary – though it still has 200 miles to go before reaching the Adriatic. The green bridge that crosses the span is over 1,000 yards long, and the river itself, rapid and powerful when melting snow streams out of Lake Maggiore in the spring, later becomes a sunny maze of channels and gravel beds, with lizards, turtles and hawks. Presiding over the confluence stands Pavia, which made it a strategically important city, as the Ticino was the road to the passes that led to northern Europe, while the Po was the gateway to the cities of northern Italy, the trading empire of Venice and the Adriatic.

Pavia also boasted the most important – and for a long time (from the fourteenth to the nineteenth century) the only – brick bridge over the Ticino, which otherwise presented an insuperable

South of Pavia, the Ticino flows into the Po before turning east to Venice.

barrier to east–west travel. As such it came to occupy a high place in the early history of this rich territory.

~

In Roman times, Pavia was known as Ticinum, and was even more dominant than Milan – as a fort, an administrative centre, a mint and a communications post. After the fall of the Roman Empire it became the capital of the Ostrogothic kingdom, then the capital of Lombard Italy, then Charlemagne's headquarters, and finally (twice) the centre of Frederick Barbarossa's Italian colony. On the marble floor of the cathedral five round stones (four black, one white) mark the exact spot where ten kings of Italy were crowned.

There are other historic echoes. The remains of St Augustine were discovered here in the seventeenth century. And it was in a Pavia prison that the English thinker Boethius wrote *The Consolation of Philosophy*, one of the great early Christian texts. In 1361, by which time Pavia had fallen under the sway of Milan's Visconti dynasty, it became home to one of Europe's oldest and most eminent universities, famous for its schools of law and medicine – Leonardo da Vinci himself studied anatomy in its halls.

But Pavia's prominent setting inevitably made it embattled too. The strategic importance of the Ticino–Po confluence, guarding both the Lombardy plain and the path up to the mountains, attracted all sorts of armies. Indeed, it was here, in November 218 BCE, that the legendary Hannibal arrived on the west bank of the Ticino, having miraculously led his Carthaginians over the Rhone* and across

---

* He forded the river near Avignon – no easy feat since it was a thousand paces wide. His army chopped down trees, lashed them into rafts, and in two days was safely over. There was one minor clash with a Roman patrol, but otherwise the

the Alps before falling on Rome itself. The Roman general standing in his way was Publius Scipio, father of the more celebrated Scipio Africanus. He had taken up residence in Piacenza, one of two Roman garrisons in the Po Valley (Rome had only recently assimilated what used to be Cisalpine Gaul), and hastened to meet the intruder at Pavia. He built a bridge over the Ticino (which later came to be known as the Ponte Cuperto or 'covered bridge') for his legions, hoping perhaps that a bravura show of force might suffice.

It didn't. Hannibal fell on him at once. It was cavalry against infantry, and the battle did not last long. A wounded Publius Scipio fled back to Piacenza, his reputation ruined. The Battle of Ticinus (named for the river) was a skirmish that rapidly became a rout. But Hannibal was not able to prevent his opponent from retreating to the safety of Piacenza; by the time he arrived he found himself camped outside a well-fortified city. Fortunately (for Hannibal had no interest in a lengthy siege) the Roman army, now under the new leadership of General Sempronius, again marched out to meet him – this time on the banks of the Po at Trebbia.

This was the day Hannibal became a name to be feared. Both sides numbered in the region of 40,000, but the Carthaginian cavalry outmanoeuvred the statuesque Romans and showed them no mercy. Only a quarter of the home army escaped back into Piacenza alive.

Thus began the Second Punic War. It started with the crossing of two rivers (Rhone and Ticino) and spilled enough blood to turn a third (Po) red. And it was early days in the tragically long history of war in Europe, much of which would play out along the same waters.

---

whole army managed to cross without a fight. It was the first victory of Hannibal's campaign, and a reminder that some of history's greatest tussles have taken place on the banks of a river.

~

After being occupied many times, by Spain, France and Habsburg Austria in turn, Pavia slowly lost its footing as the region's presiding force. And when another canal, the Naviglio Pavese, was completed in 1818 (improving Milan's access to the Po) it reduced the importance of the confluence as an interchange. Until then, cargo from the Po had been unloaded in Pavia for onward distribution; now it could be taken straight to Milan. The first steamship arrived two years later, and that too accelerated the development of Milan at Pavia's expense.

Only a few stones of the coveted Roman bridge remain now (visible when the water is low). Its replacement was built in 1354 (it was shattered by Allied bombing in the Second World War), and this medieval construction was of great importance too. The fact that it was fortified with towers and arrow slits tells us that it was a military post; but it was a cultural centre too – for hundreds of years it had a chapel in the middle. Einstein had such fond memories of his time on the Ticino that a plaque on the bridge bears a tribute from him: 'I have often thought about that beautiful bridge in Pavia.' It even inspired a popular local legend.

On Christmas Eve in the year 999 (a made-up date if ever there was one) a group of pilgrims trying to make their way to Midnight Mass were stopped by fog. A figure in red clothes appeared from nowhere and offered a typical devil's bargain. He would conjure up a bridge in exchange for the first soul to set foot on it.

The pilgrims were friends with the Archangel Michael, who happened to be in the neighbourhood and saw what was happening. He thought for a moment, accepted the deal, and then sent a goat across. It became known as the Ponte del Diavolo – the Devil's Bridge.

Does it spoil the story that this is a direct copy of the fable they tell at the other Devil's Bridge, the one painted by Turner over the gorge at Andermatt? In that version, the travellers sent a dog rather than a goat, but the effect was the same: the scarlet devil could only gnash his teeth.

There are other versions of the same tale – probably every country has one. That in itself shows how universal is the importance of a bridge – its power, its prestige and its sheer practical value as the only way across a temperamental river.

In this context it is easy to see how the Po must once have seemed as intimidating as the sea – and why a solitary bridge was worth fighting over.

16

PLAIN FOOD

Traditional maps depict Lombardy as a dark triangle cutting across the north of Italy like a bruise. The lower part of the Po, near Ferrara, is as much as 16 feet below sea level, which means that, in effect, the valley is a wide trough filled with mountain sediment. If not for that, it would be a lagoon, and the south of Italy would be a virtual island.

It is not beautiful. The Rhone and Rhine have hills, vines and ruins. The Po Valley is the opposite: hot and relentless. In 1957 the film director Mario Soldati produced a twelve-part documentary called *In Search of Genuine Foods: A Journey Along the Po Valley*, and found it unappetising. He hoped to find old men with Michelangelo faces plucking trout from silver streams, matriarchs pounding home-harvested ingredients, or children milking cows in tumbledown barns. Instead, there were factories using imported milk to process industrial-sized vats of *grana*, *mozzarella* and *mortadella* for the overseas pizza market.

In the age of mechanisation it barely feels inhabited. There might be a van in the distance, a tractor grinding or someone checking the irrigation. But not much else. It is an industrial park. When Tobias Jones prepared to paddle down the river (for his 2022 book *The Po:*

An Elegy for Italy's Longest River) he was aware of studies showing it to contain 2,642 kilograms of zinc, 1,154 kilograms of copper,
1,312 kilograms of lead, similar quantities of chrome and arsenic, and
barrowloads of cocaine as well. It was no longer anything like the
fresh water sheeting from the Alpine peaks.

Yet despite this bleak list, this is a resonant stretch of riverscape.
The humid climate and soft soil of the Po Valley – all that mountain
residue composted by the long journey south – has made it a perfect
growing medium for every sort of food: one third of Italy's food is
grown in it, which explains why the same proportion of Italy's population – a third – lives here. This wide-horizoned, river-made and
river-fed domain is one giant al fresco hothouse. Just as the Nile created the fertile heart of Egypt, so the Po has laid down and irrigated
a lush green farm in an otherwise parched world.

~

Three locales near Milan illustrate the point.

The first lies west of the city, halfway to Turin. If you cross the
Ticino on the motorway and turn north for a mile or two, you find
yourself in a treeless expanse of fields that are brown in winter, silver in spring and green in the summer. Poppies flash on the verges,
and if you stop for a breath of air you can hear the low grumbling of
frogs in the watery fields. White egrets, herons and storks pick and
jab their way through the mud.

The rice paddies stretch as far as the eye can see, rendered vivid
by the skyline – the massif of Monte Rosa, Switzerland's highest
mountain. And here we find a commune with no more than a thousand inhabitants, most of them farmers. There is little to detain
visitors – a village square, a post office, a cafe/bar, a pharmacy, a

church and a cemetery that honours the fallen of two world wars. But it is the quality of the rice grown around this village – firm, yet translucent and creamy – that makes the village synonymous with its most famous product.

The village's name is Arborio.

The rice it produces is one of many in these parts – *baldo*, *carnaroli*, *gladio*, *loto*, among others. The paddies have an Asiatic flavour – the region is sometimes known as 'Little China' – but the grains are treated with the same decorum that *viticulteurs* bring to their grapes.

The capital of this rice empire is the city of Vercelli. Of course (since this is Italy) it is an ancient place, awash with Roman and medieval relics, and home to an early Gothic abbey church (St Andrea). But its chief accomplishment is the sea of waving rice that stretches for miles beyond the car dealerships and home-improvement superstores on the ring road.* In early summer, when the seedlings are planted and the fields are drenched in a foot of water, it becomes an island.

At which point the mosquitoes take over, and no one goes for a walk.

In the Middle Ages most of the land beside the Ticino between Milan and Turin was a malarial swamp known as the *Barragio* ('uncultivated'). It was French Cistercian monks who made it productive.

It was not their first time: a central tenet of Cistercian teaching insisted that they decline tithes, which obliged them to generate income through agriculture. And they also had a rule that they

---

* The upmarket nature of the car dealerships – BMW, Audi, Alfa Romeo – testify to the fact that this is a prosperous town. It is not a humble ingredient, Arborio rice.

Anyone for risotto? The village of Arborio, below its mountains.

should wander away from their home monasteries to found sister houses elsewhere. This practice may have been devised initially as evangelism – a way to spread the word. But blended with their feeling for agriculture it became a remarkable commercial tactic.

Bernard of Clairvaux, father of the order in the twelfth century (and a co-founder of the Knights Templar as well), had left Cîteaux, his home monastery in Burgundy, a short walk from the vineyards of Nuits-Saint-Georges, to establish his own abbey at Clairvaux, north of Dijon. From there, a second generation of monks fanned out to sponsor satellite branches. It wasn't easy – they were encouraged to seek out inhospitable places where they could grow closer to God through suffering. But they were proud spirits whose peregrinations took them as far as Yorkshire, Austria, Ireland, Spain and Italy. Wherever they settled they turned wild land into productive *terroir*. Here, beside the Ticino, they drained the mud, diverted the river into a grid of irrigation ditches adroitly managed by sluice gates, and created the perfect setting for rice, an unlikely crop for the region.

The grain they grew was an import or immigrant from Moorish Spain (the easiest way to enrage a Milanese is call it Italian paella) but by the year 1700 there were some 4,000 rice farms in the fields west of Milan, and the countryside was a mirage of shimmering squares, watered by canals, with barely a tree to cast a shadow on the bright green rice shoots.

This is how this riverside paddock, Europe's largest rice farm, became home to an appetising and distinctly Italian creation: risotto.

～

The second culinary sample can be found less than an hour in the exact opposite direction: east of Milan towards Bergamo. These days it feels like a dormitory suburb, but this was once an essential centre and rest stop on the route of the 'transhumance' – the annual transfer of livestock up to and down from the mountains. A horse market in the tenth century, it lies on Leonardo's Martesana Canal, and has an ancient campanile, a well-stocked high street and a busy town square.

The surrounding terrain is featureless, and at first sight resembles the rice fields of Vercelli, but this is a land of *formaggio* (cheese).

And not just any cheese – Gorgonzola.

The fields are too hot and soggy for cattle, but the hills to the north, with lavish grass fed by Alpine water, support a dairy industry based on a seventeenth-century herd assembled when the region was ruled by the Empress Maria Theresa (Milan was then one of the jewels in the Habsburg crown).* In the creation myth, a herdsman once added too much curd before putting the churned milk to one side. The result looked awful, with its mottling of mould, but turned out to be delicious.†

It is named after the town, but its origins are not truly known. Indeed, several other places claim to be the source of this famous cheese, and the bulk of the Gorgonzola made in the region today (and it is a substantial business; it exported 2 million 'wheels' in 2023, not to mention a similar tonnage of mascarpone and dolcelatte) is produced in cavernous industrial estates around Novara, near Milan's Malpensa Airport.

---

* Maria Theresa also presided over the construction of La Scala, the greatest opera house not just of Milan but of Italy.

† This may well be an apocryphal story. Quite a number of famous foodstuffs are said to have been discovered in this accidental way (including wine) – it is what we might call a common rural myth. But in this case it does seem a plausible explanation.

In taking the name Gorgonzola, it might have been honouring the place where it was born, but it was also borrowing the glamour of Greek mythology – the Gorgon being the snake-haired temptress whose gaze turned people to stone). It was for this reason that Leopold Bloom, in James Joyce's mock-Homeric *Ulysses*, took a moment during his odyssey through Dublin to enjoy a Gorgonzola sandwich with a glass of red wine.

~

The third locale is the twelfth-century Abbey of Chiaravalle, 6 miles south of Milan's duomo. Like the church in Vercelli it was founded by Cistercians, and in this instance Bernard of Clairvaux was personally involved. Chiaravalle was a gift from the pope in 1135 – a reward for going to Milan to support Innocent II in a political row. It was then a parcel of marshy land on the road to Pavia, but with Cistercian help it soon became a thriving monastery farm.

The land was turned into a dairy farm that produced a hard, Parmesan-like cheese that was allowed to mature and intensify for a year. The Cistercians called it Gran Padano, after the place where it was born: the name literally means 'granular from the Po valley'.

The twelfth century did not bother to patent places as 'protected designations', so in due course this Cistercian Italian cheese became generic – unlike Parmesan, which had to hail from Parma or Reggiano, a hundred miles down the Po.* Today, it is produced all over the world.

---

* Interestingly, the famous lothario Giacomo Casanova wrote in his memoirs that Parmesan was not a genuine child of Parma but in fact was made in Lodi, near Milan. Perhaps he was getting it muddled up with the Cistercian *grana*.

~

These are just three of the appetising treats rooted in the hot, flat land beside the Ticino and Po. There are others we could name: salami milano, the beef, pork and pepper sausage that is a staple of every delicatessen in Europe, or ossobuco, a Milanese veal shank (the name means 'bone with a hole') served with yellow saffron risotto. Taleggio and mascarpone come out of the pastures of Bergamo and Lodi. And panettone is thought to have been invented here, too, as a way of using up Milan's leftover fruit. The earliest mention of this bread-cake comes from Pavia; the modern world makes a hundred million loaves a year.

Further down the Po we find Parma ham, polenta, maize and wheat – this is Italy's bread-and-pasta basket. No wonder people were hooked when a famous BBC *Panorama* April Fool in 1957, which purported to show the spaghetti 'harvest' being picked in pasta orchards, referred solemnly to the 'vast spaghetti farms in the Po valley'. The plain provides flax, hemp and every sort of vegetable, while the lakes and the rivers offer fish and shrimp. The river itself is a slow-moving giant, oozing in the heat, and some of its fruits are monstrous: carp the size of sheep, catfish the size of calves. In the summer of 2023, when the water was low, a fisherman named Allessandro Biancardi hooked a giant catfish that was officially described as 'prehistoric'.

None of this would be possible were it not for the happy marriage of Alpine water and Mediterranean sunshine. Naples has its pizza; the Amalfi coast and Sicily have citrus fruit; focaccia is Tuscan, minestrone Roman. But everything else we think of as belonging to *cucina Italiana* can be found in the Po Valley. It is where the mountains meet the sun, all watered by the river.

17

ROME-SUR-RHONE

When the Rhone leaves Lake Geneva its flow is controlled by the Seujet Dam, a modern version (constructed in 1995) of the old Coulouvrenière Barrage. In keeping with best modern practice, it fulfils its ecological obligations by having a 'fish ladder' and a 'beaver ramp', but its primary purpose is to manage the river's flow while also supplying Geneva with energy.

Released from those constraints, the river then bounds through the Jura and on into southern France. When Victor Hugo called it 'a tiger', he was probably thinking of this vigorous upper section. Boosted by the Arve, which meets it in Geneva, rich in debris from Mont Blanc, it becomes a cutting tool strong enough to tear through the hills around Lac de Bourget (France's deepest lake, with Aix-les-Bains as its thermal spa), then on to the union with the Saone at Lyon – capital of the confluence.

Lyon is France's third largest conurbation, after Paris and Marseille, meaning that two of France's greatest cities lie on or near the Rhone. Naturally it is the centre of the Rhone-Alpes region, full of imposing buildings that spill across its two rivers and over an abrupt hill. Its location alone made it an important site from its earliest days. It sat in the heart of an arable paradise, with the wine

148

country of Burgundy and Beaujolais to its north, and was also close to Strasbourg, Basel and the cathedral cities of the Rhine.

But the river was its gateway to the world. Two momentous cultural influences entered France via the Rhone, with reverberations that would underpin the entire region's history.

The first of these was Rome, because 2,000 years ago this was the centre of Roman Gaul: a great colony that made Lyon (known then as Lugudnum) its headquarters. This transforming conquest approached France from the south. In those days the mountainous coast between Italy and France (the Côte d'Azur, where Monaco and Nice now face the blue sea) was too rugged to be passable, so the Rhone was the only way to the interior. As the invading force made its way up the valley it found a familiar landscape, all blinding sunshine and gnarled olives, very like Rome itself. It established a splendid new province (literally: Provence) with imposing cities at Arles, Nîmes, Orange and Vaison-la-Romaine, and though the going was rough – the Celtic Gaullois fought for their lives; perhaps as many as 100,000 legionnaires died at Orange – it planted its urban culture wherever it settled: roads, walls, aqueducts, theatres, baths and temples. Pushing north towards the milder climate, it made its capital where the rivers met – a place that was both easy to reach and easy to defend.

Lyon grew behind tall ramparts on the hill above the near-island, with a population numbering anything between 100,000 and 200,000. Soon it had two huge amphitheatres, both with a view of the Alps. The ruins are still there, a brisk walk uphill from the city centre.

Rome's presence in Gaul was hardly temporary. After Julius Caesar subdued the Celtic-era tribes in 50 BCE, Rome went on to rule the 'Three Gauls' (Belgium, France and Switzerland) for roughly 500 years, with Lyon as its capital. France became thoroughly Latinised in language, religion, architecture (Romanesque), civil arrangements and artistic leanings – everything from the Pont du Gard and the Latin Quarter to Asterix and villas by the sea. When Rome eventually withdrew, in the fifth century, it could not help but leave an indelible part of itself.

~

The second momentous thing that reached France up the Rhone was Christianity. According to legend, the Mediterranean town of Saintes-Maries-de-la-Mer, next to the mouth of the Rhone, is named after Mary Magdalene, said to have landed here after fleeing the Holy Land (accompanied by Lazarus).* And while this may be a fanciful story, the Rhone was certainly the route along which the new faith arrived. Monks and missionaries from Italy, and refugees from further afield, all made their way up the valley that led to France's mysterious interior.

The stones behind Lyon's cathedral claim to be vestiges of a church built in 150 CE, making it the second-oldest Christian relic in Europe. Those early Christians in Gaul were Greek rather than Roman – until Emperors Constantine and Licinius issued the Edict of Milan in 313 CE, which decreed religious tolerance, they were persecuted by Rome. But they advanced into Gaul via the Roman road, like everyone else.

Little is known of that first community in Lyon aside from the fact that in 177 CE it was the stage for one of Rome's most violent reprisals against the new Christian faith, with many arrested and killed in

---

* This legend featured in *The Da Vinci Code* as part of the great Church cover-up.

what became known as the Persecution of Lyon. Various legends of martyrdom were born from that time. According to one, forty-eight Christians, led by Pothinus, a ninety-year-old Greek from Smyrna and the first bishop of Lyon, were thrown to the arena's lions. And another told of a slave girl named Blandine who after refusing to recant her faith was tied to a cross and delivered to the prowling cats – which responded to her fervent prayers by leaving her untouched.

Rome was then at its mightiest extent – easily strong enough to suppress passing heresies of this sort. But these early expressions of Christianity were the advance guard of a religious upheaval that in time would undermine Roman authority, and then infiltrate and alter it.

Roman Christianity went on to become an enduring part of France's culture. Its cathedrals – Notre Dame, Chartres, Rouen, and a hundred others – are one of the world's most admired architectural treasures (and visitor attractions) to this day. The Rhone, meanwhile, became rather more than a Roman road: it became the headquarters of the Catholic Church and, in effect, an overseas branch or colony of the Vatican.

This part of its life began in 1309, when Rome was writhing in civil conflict. The Vatican had chosen the French Archbishop of Bordeaux, Bertrand de Got, as the new pope. Taking the name Clement V, he stayed in France for safety, and his successor Jacques Duèse (Pope John XXII) did the same. Avignon, a Roman stronghold with a bridge over the Rhone, became the seat of the papacy. The following seven popes, five of them French, ruled in style from Avignon's princely Palais des Papes.

It meant that France suddenly had an unusual presence – effectively a foreign city state – in its midst, planted on the old Roman river. It was known as the Comtat Venaissin, and was given to the

Holy See by the Count of Poitiers and Count of Toulouse in 1271. The Roman Empire had marched up the Rhone to Lyon; now the new Christian empire was taking exactly the same route to carve out its own province in Provence.

The religious capital was Avignon, but the administrative centre was Carpentras, a Roman town of arches, columns, gates, aqueducts, libraries (and truffles). This was where Clement V first took refuge, and papal rule went on to make the Comtat a place of special privileges, with citizens exempt from military service and French taxation (naturally, the Church was happy to accept their tithes instead). When France turned against its Jews in the fourteenth century, the Comtat offered them a home. It cannot be said that this was thanks only to papal virtue: the Jews had to submit tax to the bishops in return for their protection. But it did lead to Carpentras, a Catholic stronghold,* also becoming the site of France's oldest synagogue in 1367.

The town retained its aura of miraculous destiny when it became one of the few to be untouched by the Black Death. So when the papacy returned to Rome in 1376/7 the Comtat Venaissin remained a favoured enclave. It even managed to resist repeated French assaults until the Revolution swallowed it up in 1791. The great walled Palais des Papes in Avignon became a military barracks, and the once mighty Comtat Venaissin became an ordinary French *département*.

It is one of Europe's forgotten princedoms, appearing only as an odd-coloured patch on antique maps. Few holidaymakers in Provence have even an inkling that for so many centuries (1274–1791) this county-sized strip of Vaucluse was once an embattled papal province. It is hard to imagine England permitting, say, Oxfordshire

---

* The cathedral in Carpentras contains (supposedly) a nail from the true cross.

Lyon: where the Rhone twists south to Avignon and the Mediterranean.

to be an overseas colony for so long, or being able to forget about it so quickly.

But while it is no longer a tangible domain, traces of the old Comtat still stick up like the spars and masts of a sunken ship – a ruined wall here, a monastery door there, a lump of old fortress hidden in the pines. And since the popes were powerful patrons of the arts, there were other consequences. It was thanks to Avignon that France was introduced to Italian sculptors and artists such as Giotto and Duccio.

In this way the Rhone has always been more than a pleasant ribbon curling and furling its way through the *paysage* south of Lyon. As the primary route into France for two extraordinary cultural traditions, it has been a central force. Plaited together, classical Rome and Christianity wove themselves deep into all aspects of French life: its religious and philosophical ideas, architecture, tastes (food and wine), financial habits, social structures, legal principles and artistic leanings. Pretty much everything we now think of as recognisably French has origins in the world that sailed out of the Mediterranean and moved up the Rhone.

~

In the succeeding centuries Lyon continued to be a religious rather than a commercial centre, the seat of a senior archbishopric and home, thanks to its location, to some notable abbeys. But in later eras its ideal setting on the meeting point of the rivers made it a silk-working and financial metropolis. The river was fast-flowing, complicated and seasonal, but navigable from ancient times onwards – even if it did require large teams of horses to drag the barges upriver. So when, in 1540, Francis I gave Lyon the monopoly for all the silk coming into France (up the Rhone) from Asia and Italy, he turned the city into a manufacturing base; by the nineteenth century there were as many as 100,000 looms spinning in *vieux Lyon*, with narrow alleys connecting them to the river and its boats.

The wealth this generated helped Lyon become an important centre of printing, second only to Paris, which gave it a leading voice in France's intellectual life. The slim peninsula between the streams, the so-called *Presqu'île* ('the almost island') is these days a bustle of historic lanes, courtyards and civic halls – *mairie, préfecture*, opera house, library, law courts. As one end of the historic Silk Road between Venice and China via Samarkand, its riverside location opened it up to the whole world.

The fertile river valleys also placed Lyon at the vanguard of French cuisine. Sitting between the mountains and the plain, and with easy access to both land and water, this was the mouthwatering meeting point of north and south – butter and cream one way, fat tomatoes, olive oil and lemons the other. It contained the finest larder known to humankind – fish from the rivers, beef from Charolais, every variety of grain, fowl, milk and cheese from the nearby pastures, truffles and game from the woods, along with bustling cartloads of charcuterie, fruit, nuts, vegetables, bread, wine and everything else a chef might desire. Soon it was also being said that in fact there were three rivers flowing through Lyon: the Rhone, the Saone – and Beaujolais. That was a way of recognising that Lyon was *the* food capital or 'belly' of France.

Not that the origins of Lyonnais cuisine were ambitious. If anything, they leaned on low-born ingredients such as offal, tripe and other innards – peasant foodstuffs. Lyon's reputation as the home of haute cuisine was actually based on the domestic cookery of the so-called *Mères de Lyon*: barnyard fare enriched by butter, cream and wine; stewed eels, pâtés and mousses, truffled chicken and pastries. Centuries earlier, Erasmus himself had once commented favourably on the nourishment produced by 'the Lyonnaise mother'. By the twentieth century, there were hundreds of small restaurants or *bouchons* in the city, all called *Mère* this or *Mère* that.

The most celebrated was the one run by Eugénie Brazier. Brought up on a farm outside Lyon, she opened her first restaurant (*La Mère Brazier*) in 1921 and launched another up in the hills soon afterwards. Fortune was smiling: the *Guide Michelin*, launched in 1900 as a lunch-planner for motorists, was just then bringing in its three-star rating, and both of her restaurants were so garlanded. Brazier became the first chef ever to win six Michelin stars, and though she never disguised her ordinary roots ('I learned to cook by doing it') she attracted an elite clientele. Edith Piaf was a frequent visitor, as was Lyon's mayor, Édouard Herriot (three times prime minister of France, his name lives on in the city's port). He claimed that Eugénie had done more for Lyon's prestige than he ever had. But she turned down the *Légion d'honneur* on the grounds that so grand an award should be reserved for things 'more important than cooking'.

Lyon has a dark side too. It was the scene of massacres in the Wars of Religion against the Huguenots (Protestantism having sailed to Lyon down the Rhone from Calvin's Geneva). The French Revolution arrived here after a lengthy siege; and in the Second World War, when the city was occupied by Nazi Germany, it was the headquarters of the French Resistance, full of secret newspapers and underground plots. As a result, it was also the home of the Gestapo – indeed it was in a converted school building a few yards from the Rhone that Klaus Barbie, the 'Butcher of Lyon', did his dirtiest work. German Shepherd dogs were trained to maul naked women, and Jean Moulin, leader-hero of the Resistance, had his fingernails torn out, perhaps by Barbie himself.* That school has since been turned into a museum of the French Resistance and is now a shrine to the 80,000 Jews deported from Lyon in those years.

But even these horrors cannot quite be said to have had an impact as profound as the earlier imports that the river brought to this part of the world. The rich culture and language of ancient Rome, and the heady emotional drama of Roman Christianity – these were nation-building influences.

And there was another civilisational treasure escorted into France by the Rhone, something every bit as integral as the others.

Wine.

---

* When Barbie asked Moulin to write down the names of his fellow conspirators, the prisoner simply drew a cartoon of his interrogator. Barbie was later (much later – in 1983) tried in Lyon's Supreme Court, facing the Rhone near the cathedral, found guilty, and executed.

18

HISTORY BOTTLED

The first thing you see when you walk into Vienne's Office du Tourisme is a towering display of some 900 bottles, all produced nearby. For 2,000 years this has been the capital of a famous wine region: Côte-Rôtie. A few miles south of Vienne the river pours into the gap formed when prehistoric Alpine rock crumpled into the Massif Central and created a chute. That was in Cenozoic times, roughly 65 million years ago, but it set in stone the Rhone's future course.

This miniature city 20 miles downriver from Lyon was founded by Julius Caesar himself, and feels even more obviously a Roman colony. It takes only ten minutes to walk from the cavernous amphitheatre (there are two, but only one has been excavated) down through the Gardens of Cybèle and the Roman-era Forum to the Corinthian Temple of Augustus in the town centre.* Thomas Jefferson was so impressed by its classical columns, when he visited Vienne in 1787, that he echoed it in the public buildings

---

* When Rome left, this grand building became a parish church, and after the French Revolution was secularised into a 'temple of reason', a law court and a library.

of Washington. There are age-old ramparts on all sides, and the vestiges of a once opulent Roman villa watch over the other side of the river.

Its ancient heritage was neglected until its long-buried bones were unearthed by archaeologists in the 1960s. But the theatre is on the same scale as the one in Lyon (capacity: 11,000) and now stages summer jazz festivals. A fragment of the original Roman Circus in a traffic island – a stone needle known as the *Pyramide* – gave its name to one of France's most famous restaurants, founded in 1924 by the Michelin-starred chef Fernand Point, one of the founding fathers of French cuisine.

Vienne likes to claim that Pontius Pilate took his last breath here – the town boasts a *Maison de Pilate* and a column in his honour. Sadly, there are towns in Germany, Spain and even Scotland that make similar assertions, and the mountain that rears over Lucerne is named Pilatus on the grounds that he drowned in the lake up there. But with sun glinting on these 2,000-year-old stones, Vienne does indeed seem like exactly the sort of place to which a judge might retire after a case as controversial as the one that made Pilate wash his hands.

As it had at Lyon, Christianity thrived in this Roman-tilled soil – especially after Emperor Constantine made it the official state religion in 325 CE. Vienne's cathedral (part Romanesque, part Gothic) was dedicated to St Maurice, the martyr who refused to kill Christians on the Rhone above Lake Geneva. It took several hundred years for news of his heroism to drift down to Vienne, but his head eventually arrived as a saintly relic in the eighth century. Five hundred years later, in 1311, the Papal Council that abolished the Knights Templar was held right here in Vienne. By this time it was a centre of faith as well as a centre of power.

Vienne: capital of the Roman Rhone.

It is hard not to feel a quickening of the pulse as one enters the pass through the hills in which the city reclines. The outskirts of Lyon are as grubby as any urban back office: mile after mile of oil refinery, power station, container park, railway shed, gas silo, sewage plant, gravel pit and timber yard. But soon the high ground on the right cramps in so close that the roads and railway lines have no option but to squeeze in tight like the nerves in a hip joint. The exposed cliff becomes a near-vertical incline perfectly angled to receive the morning sun.

The Romans looked at that cliff and knew at once what it was for. They cut narrow, perilous terraces into the rock, and planted vines.

〜

The historian Thucydides went so far as to attest that Europe 'emerged from barbarism' only after it had learned to harvest the olive and the grape – the fruits of all wisdom. And indeed it would be a mistake to think of wine as a trivial or marginal component of European civilisation, for few things have played so emphatic a role. Robert Louis Stevenson called wine 'bottled poetry'; Louis Pasteur insisted that it contained 'more philosophy than all the books in the world'; Jefferson called it 'a necessity'. 'In vino veritas,' wrote Pliny the Elder, while Napoleon was among thousands who saw it as an indispensable companion to both happiness and misery: 'In victory you deserve champagne,' he said, 'in defeat you need it'. Galileo summed up wine's earthly qualities by calling it 'water held together by sunshine'.

These are secular plaudits, and it is true that wine has been embedded in Europe's social texture for thousands of years. Plato's symposium was essentially a licentious wine-tasting club (with the drink watered down, of course; only barbarians drank their wine neat in those days). In fact it is not originally European: fossilised pips dating back 7,000 years have been found in Lebanon, Jordan and Syria; the earliest known vineyard is in Armenia; and one of the oldest creation myths is Persian. In this story, a slave in the royal harem tried to end her days by poisoning herself with a jar of rotting fruit – only to find that it made her deliriously happy.

It may have been Greeks who planted vines in this landscape – they did have a colony at Massalia (Marseille), and archaeologists have found amphora fragments. It might equally have been indigenous Gauls. But there is stronger evidence, in plentiful coin hoards and pottery shards, to suggest that it was the Roman founders of Lyon and Vienne who first hacked significant terraces into the walls of this valley, and made wine a definitive French product.* The geographer Pliny the Elder mentioned the wines of Vienne as being expensive in Rome, which tells us that Côte-Rôtie was internationally renowned even then.

Rome went on to plant vineyards all over the country, from Provence to Champagne. Bordeaux, Burgundy, Alsace, Savoy, Loire, Languedoc . . . different soils, different weather, different grapes – yet all of them indefinably French.

Britain played a role in this as a land of *very* enthusiastic drinkers, but without its own supply. When Eleanor of Aquitaine married Henry II in 1154, she cemented an alliance that made

---

* The museum at the Château de Guigal has an impressive display, including a Greek silver drachma from Marseille in the second century BCE, and bronze *sestertii* bearing the heads of Julius Caesar, Marcus Aurelius, Tiberius and Agrippa.

the claret of Bordeaux a fixture in British life, and helped France become the world's most revered wine economy. The broader culture this created – all those abbeys, *vignobles*, *caves* and chateaux – is an entrenched part of what France calls its *patrimoine* – its national inheritance.

By this stage wine was already more than a social mood-enhancer. Indeed, it is as a religious emblem that it became Europe's quintessential drink. When Christianity emerged, it adapted pagan-era stories: the wine that pulsed through the veins of Bacchus became the blood of Christ. The famous miracle at the wedding in Cana, when Jesus turned jugs of plain water into marriage wine, somehow told the entire Bible story in miniature – Old Testament water was magically transformed into the heady possibility of redemption. And then, at the Last Supper, bread and wine were transfigured into the body and blood of Christ himself.

Vineyards thus became uniquely important to the medieval Church. Some of France's most celebrated names were planted by religious foundations – Dom Pérignon, who may not actually have invented champagne but was certainly a distinguished figure in its discovery, was a seventeenth-century Benedictine monk.* When Bernard of Clairvaux decreed that his monasteries create networks of sister houses and labour in the fields, he was unwittingly creating agribusiness. And the first fruits of this venture were the historic wines of Burgundy.

The monastic taste for wine expanded into a mainstream enthusiasm. Yet even today, though faith has retreated and there are vines all over the world, it remains one of the ways in which Europe knows itself. Mealtimes, social ceremonies and religious services follow

---

* If anything, he was looking for ways to get the bubbles *out* of champagne.

much the same pattern from Ireland to Poland, and from Norway to Sicily.

Nor can we ignore wine's commercial importance. It is estimated that half a dozen champagne corks pop somewhere in the world every second, and that is only the outward fizz of the industry that sailed into France up the Rhone. It sailed down the Rhine too. According to Hugh Johnson (in *The Story of Wine*), as far back as the fifteenth century 100 million litres were passing through Strasbourg every year. Today, Europe produces more than 16 *billion* litres per annum. France's share of that trade is worth in the region of £13 billion, with Côtes du Rhone alone filling the equivalent of 100 Olympic swimming pools.

It is no longer a small business. And wine has fostered a hundred ancillary crafts as well: not just in barrel, cork and glass-making, but in distribution, finance, law, hospitality and retailing. So, while it is a viticultural cliché to say that wine is 'history bottled', the vineyards of the northern Rhone are indeed a place where the past feels ever present.

~

The Rhone took Rome's vines all the way to Burgundy. And it created the soil conditions that allowed them to flourish. The white pebbles that line the terraces at St Joseph, for instance, are Alpine rock, abraded and polished by the journey south. They make an ideal top dressing for the precious grapes, keeping the earth cool by day and warm by night.

It is almost as if nature knew all along what it had in mind.

The writer Jan Morris once poked fun at the solemnity of such talk when she interrupted a drive through France to pay her respects to one of Burgundy's – and the world's – most famous drinks. 'For the first and probably the last time in my life,' she wrote, she strode into a shop and bought a 'Grand Cru Montrachet – Marquis de Laguiche, vintage 1993.'

Then, asking a 'kindly waitress' to uncork it, she picked up 'a hefty ham and cheese baguette,' and asked a passing wine merchant to direct her to the exact spot from which the wine came. He pointed the way, and a few minutes later Morris was sitting on a low wall among the vines.

'There I sat, and drank, out of a plastic mug, the most famous dry white wine on Earth. It was very peaceful, rather like picnicking in an extremely upmarket cemetery . . . The wine, of course, was incomparable.'

In homage to her example, I decided to do the same thing with the undisputed star of the Rhone, Côte-Rôtie.

The English translation – 'sun-baked slopes' – accurately describes the feature that makes the Côte-Rôtie so well made. This is where the Rhone, driving south-west, chopped a cleft into the edge of the Massif Central. The result is a south-facing hill that rises a kilometre above the villages of Ampuis and Condrieu, south of Vienne. Too steep and stony for any other crop, it is fully exposed to the sun while having enough height to catch the breeze. As elsewhere, the rocky slopes act as night-storage heaters, protecting the grapes from the extremes of heat or frost. The vines, meanwhile, with their 80-foot roots, are happy to lick up moisture from the cold interior, and love the quick-draining surface.

The Côte-Rôtie terraces were nearly wiped out by the vine-eating grub *phylloxera* at the end of the nineteenth century, and the scale of the death toll in the First World War (the mural at Vienne station honours the 763 young men who died) made it too expensive to cultivate this intimidating land, which was too precarious for horses or even oxen to be of any help.

Côte-Rôtie faded from view. But in the 1970s a winemaker named Étienne Guigal and his son Marcel resolved to revive the *terroir*, and turned it into one of the world's most estimable *appellations*.*

The first bottle I looked at in Ampuis cost £400, which made me think twice (at least) about the idea. But the kindly lady in the *caveau* opposite the church was delighted to uncork a less exclusive bottle, and show us on the map exactly where it came from. Pausing only to collect a little something from the *boulangerie* on the Place de l'Église, we walked up the well-signposted Sentier des Vignes, sat on a sunny stone wall overlooking the slope where the wine was grown, and took a sip.

It was nothing like sitting in a cemetery, even an upmarket one. It was October, so the grapes had recently been harvested, and the ranks of vines looked as dazed as a regiment of new recruits standing to attention (they are well trained) at their posts. The river was a flat grey streak on the far side of Ampuis, further away than I had expected – it didn't seem likely that it could have any direct impact on the grapes. But the indirect impact was beyond debate. The river had created this entire habitat.

It was very quiet, and extremely beautiful. Looking down the valley we could see nothing but the flowing sweep of the Rhone, the green cliff cloaked in vines, and, in the distance, the white dome and wisp of steam hissing out of the Saint-Alban nuclear power station beyond Condrieu.

---

* Robert Parker's encyclopaedic *Wines of the Rhone Valley* (1997) finds Guigal's 1995 Côte-Rôtie La Mouline 'so gorgeous and hedonistic that I often wonder why it has not been declared illegal'.

The vineyards of Côte-Rôtie lord it over the fast-flowing river.

The wine (of course) was extraordinary. And I know how important it is to have exactly the right kind of wafer-thin crystal in order
to maximise one's sense of colour and bouquet . . . but it was pretty
good straight from the bottle, too. The earth was damp; the sky was
blue. The leaves protecting next year's grapes were preparing to go
again next spring.

Were we imagining it, or was there a flash of ancient Rome in
there, plus a hint of medieval sanctity, of candles and Latin chant?
That is the point of wine – it expresses land, but also time. I recalled
what Hugh Johnson wrote in 1961 when he tasted a 1540 German
Steinwein. Exposure to modern air was rapidly turning it to vinegar,
but 'for perhaps a few mouthfuls we sipped a substance that had
lived for over four centuries . . . It even hinted, though it's hard to
say how, of its German origins.'

The origins in *our* mouths were, like the language we spoke, a
time-honoured compound of Greek, Italian and French ingredients. In other words . . . European. And for that we had the river to
thank. The valley that stretched away to the south below our feet
was responsible for much more than this one exceptional vintage:
it had created the primary route by which France received three
things without which it would hardly be France: Roman civilisation,
Christianity and Côtes du Rhone.

~

Côte-Rôtie is not the region's only celebrity. As the river approaches
Tournon-sur-Rhone it skirts a geological outlier: a lump of Massif
Central that has been stranded on the wrong (eastern) side of the
waterway. It is a notable landmark, a tall dome visible from miles
around, but also more than that. The dry soil, southerly aspect,

hot sun and refreshing wind make this another remarkable terrain for wine.

The Colline de l'Hermitage (usually translated as 'Hill of Hermitage', but perhaps more naturally rendered as 'Chapel Hill') contains nothing but grapes. As on the Côte-Rôtie, the vines are angled to drink in all of the available sunshine while the hills catch the rain. And this bend in the river has the same fortunate combination of qualities: it lies in the hot latitudes of Provence, but has enough height to enjoy cool evenings.

It is textbook wine scenery. Up in the river's fast-flowing Swiss incarnation, the Rhone produces fresh vintages tinted with an echo of mountaintops. Down south it has made a vast bowl for hot-country reds.* But in between, on the monumental slopes of the Colline de l'Hermitage, as at Côte-Rôtie, it produces one of the planet's most sought-after drinks.

The hill takes its name from a thirteenth-century knight, Henri Gaspard de Sterimberg, who (so the story goes) returned from the Albigensian Crusade, the campaign against the Cathar heretics of south-west France, and retreated to the penitential chapel or hermitage he built on the hilltop looking down on the river. There he lived in reflective solitude, nursing his wounds and minding the vines that clung to the slope at his feet.†

He was walking in Roman footprints. The occupiers had left a temple to Hercules on de Sterimberg's hill, an apt tribute to the labours required to cultivate vines on these daunting slopes. The temple is long gone, and the medieval chapel had to be rebuilt in

---

* Officially listed, in 2023, as amounting to 465 million bottles – quite a big party.

† The house of Jaboulet still produces a white wine – the Chevalier de Sterimberg – whose label features an image of this crusader and fetches around £100 per bottle.

1864. But though these days it stands next to a mobile-phone mast, the magic remains. And when the papacy relocated to Avignon in the fourteenth century, it put the Church's power behind an expansion of wine production. When Thomas Jefferson visited in 1787, before the Revolution, he wrote in his diary that it was 'the first wine in the world without a single exception', and bought 550 bottles.

In the age of social media, few politicians would risk such a spree. But the slopes above Tain-l'Hermitage have been legendary ever since.

The same goes for the pope's 'new castle' at Châteauneuf-du-Pape. Originally named Castro Novo, after its Italian proprietors, this hilltop palace became world famous as a wine – the crossed keys emblazoned on its bottles announce it as a premium papal vintage, and a bottle of the 1995 Château Rayas will cost you the best part of £1,000. The town also has a contemporary claim to fame as being the only one in France, perhaps in the world, where UFOs are officially banned. This came to pass in 1953, when the movie version of H. G. Wells's *The War of the Worlds* was released in France. The original 1898 adventure yarn about alien invasion had made America swoon when a radio dramatisation by Orson Welles lulled audiences into believing it was a news bulletin, and the screening of *La Guerre des mondes* led to a similar increase in alien sightings. One, gleefully reported in the *New York Times*, involved a labourer from Châteauneuf-du-Pape who spotted 'two Martian visitors' on the railway tracks outside his house, a 'cigar-like machine' and a blast of green light.

It seemed fitting that this should happen on the banks of the Rhone. It is a river of visions, marinaded in Provençal poetry and the blazing landscapes of Vincent van Gogh. But the publicity-minded mayor decided to issue a prosaic ban. 'The overflight, landing and take-off of any aircraft,' read his straight-faced decree, 'known as flying saucers or flying cigars, of any nationality whatsoever, are prohibited on the territory of this commune . . . any aircraft will be immediately impounded.'

It could be argued that the ban worked: no such craft, or person, has ever been jailed. So, in 2016, it was reimposed, allowing wine connoisseurs to detect a soupçon of distant galaxy in the famous vintage. One Californian winery actually produced a wine named 'Le Cigare Volant'.

~

Côtes du Rhone might be the most celebrated of the agricultural delicacies sustained by the river as it cuts through the south of France, but it is by no means the only one. The Rhone-Alpes region is another area of unbelievable fertility, bursting with fruit, vegetables, nuts, fish, meat, olives and tropical plants. Ever since 1922, the town of Tain-l'Hermitage has been home to a refined chocolate bar, Valrhona. Initially named the Chocolaterie du Vivarais, after the war it was rebranded to honour its birthplace on the river: Valrhona means 'valley of the Rhone'. As if mimicking the wine culture on the surrounding hills, its elegant and refined cocoa bar was the chocoholic's equivalent of a *grand cru*.

The confectionary heritage of Montélimar, a little down the river, goes back further. The flat, alluvial surroundings, richly spread with silt washed out of the mountains, mean that this part of the valley was not wine country. But for centuries Montélimar has been synonymous with the almond-honey mousse known as nougat (from the Latin *nux gatum*, or nut cake) for centuries. The perfume-producing lavender fields in these parts were a paradise for bees, whose warm, scented honey, whisked up with egg whites, and dried in the hot sun, made for a luscious mouthful. The honey-pistachio palette suggests a Moorish influence, which also would have made its way inland from Marseille, up the Rhone.

Nougat became a more-than-local delicacy when the Paris–Marseille railway arrived in 1849. Montélimar became a convenient place to break the journey, and pedlars gathered in the *place* outside the *gare* to sell their nutty pick-me-ups. Things took a mass-market turn when the Route Nationale (N7) opened in 1968, following the railway to the coast. The town still produces something around 3,000 tonnes of nougat per year, in a rainbow of flavours.

The arrival of high-speed transit has left some of the valley's other towns feeling somewhat abandoned. Picture-book Tournus-sur-Rhone, directly opposite Tain-l'Hermitage, was once a busy wine port (and thus home to the oldest secondary school in France, established in 1536); now it is mainly a road sign flashing up on the autoroute. In much the same way, Valence, once the Roman 'gateway' to the south, is now one of those stations on the TGV line where hardly anyone gets off. There is a stream of tourists on the water – nearly 17,000 passengers a year cruise this stretch of the Rhone as it rolls on to Orange (with its superbly preserved Roman amphitheatre) but the river has largely been supplanted by faster and more modern methods of travel.

That has not done much to change the pace of life in Provence itself. It continues to feel like a colonnaded piece of Roman Mediterranean, roasted by ancient history as well as dazzling sunshine. As Lawrence Durrell wrote, 'Caesar's vast ghost' continues to haunt its chapels, markets, village squares, olive groves and wine terraces. As a result, for 2,000 years it has been 'a sort of laboratory in which the European sensibility was perpetually trying to forge itself'.

Hot and fragrant, the spirit that shines through its shimmering lavender fields strikes many (British travellers especially) as expressing something profound and essential about France – a villa in Provence remains a cherished dream. And a large part of that is based on its famous wines. Even now, no English picnic can call itself complete without a glass of ice-cold Provençal rosé.

19

THE SECOND COMING

Many reasons have been advanced to explain the collapse of Rome's empire. Sheer geographical overextension, the unstoppable eruption of the German tribes, internal decadence at the centre, the loss of cohesion caused by immigration, the economic cost of so much war and conquest, the spiritual cracks opened up by the eastern empire in Byzantium, Rome's own mistreatment of its subject peoples, the gentle rise of Christianity ... These are only the headline possibilities.

But there is another, simpler explanation. On 31 December 406 CE, the Rhine froze over and became, for a few days at least, easy to cross.

An unimaginable blizzard of Germanic peoples, already starved by the icy winter, rampaged across the river and streamed on in search of food, vengeance, freedom, sunshine and everything else they lacked. Mainz, Worms, Speyer and Strasbourg were the first to suffer (interestingly, Cologne was spared – this assault was a *southerly* advance). But the storm raged over the walls and parapets of the Roman Rhineland, and eventually (in 476 CE) made its way to the portals of Rome itself.

That is how the empire fell: as if swamped by a river that was bursting its banks. In a gigantic and unstoppable migration, the immense German tribes (Visigoths, Alemanni and Franks), all being pushed west by the advancing Huns (the Romans called them all 'barbarians'), raged out of their forests, swirled across the Rhine, tore through Romanised Gaul and on into Italy. The result was a new realm, Lotharingia, a Franco-German compound that ran all the way up the Rhine and down the Rhone and Po.

Nothing could so clearly demonstrate the way that rivers change the course of history. Sometimes they are lines of contact and communication that need to be kept open; at other times they are barriers; they can be prizes worth fighting for or hazards to avoid. But in any of these guises, they are instrumental. Because even if the above reasons for Rome's fall are true, there remains the pivotal contribution of the Rhine. It was the Great Wall of Roman Germany, and though the Romans could hold any bridge, no one could survive an advance on so broad and furious a front. After centuries of ruthless dominion, Rome itself was brushed aside.

The end.

The trouble with this story is that it might not be true. There is no solid evidence that the Rhine froze at that time. The idea derives from a famous passage in Edward Gibbon's *Decline and Fall of the Roman Empire*, in which the author infers that the river was 'most probably frozen' – on the grounds that those axe-wielding tribes could not conceivably have crossed otherwise. It is a thesis not a fact.

Of course, the Rhine *can* freeze. There are allusions to such events in classical literature – it turned to ice in 366 CE, allowing the Alemanni to romp through Strasbourg and Alsace. And in 1929 (there are photographs) the Rhine froze solid once more. This was at Ludwigshafen, a major port, and it followed a week in which temperatures plummeted to −21°C. Factory workers could take their lunch out onto the Rhine – or, if they happened to live on the far side, they could stroll over from Mannheim and back across a firm pavement of ice.

Whether the great freeze of New Year's Eve, 406 CE ever happened, we cannot say. It *might* be true: the German tribes definitely did cross at this time, so it does make logical sense. But isn't that date a touch too conveniently fairy tale? A river freezes, a dam breaks, and abracadabra, it is Happy New Year all the way down to the Tiber . . .

However it happened, in 476 CE Rome fell to the Goths, and that was the end of the western part of the Roman Empire.*

~

Rome did not die completely, however. It re-emerged in one of its own colonies, the Rhineland, in a new form: as the Holy Roman Empire of Carolus Magnus, or Charlemagne. It was a religious as well as a secular power, and while its theological roots lay in Italy, its political strength was in Aachen and Cologne. It was founded on Christmas Day 800 CE and for the following thousand years, until Napoleon dissolved it in 1806, it was Europe's supreme authority. An empire that had been born on the banks of one river – the Tiber – had been reborn on another – the Rhine.

Trier was the head office of the new empire – the 'Rome of the North'. Located a few miles up the Moselle from the Rhine confluence at Koblenz, the city grew out of a Roman settlement, Augusta

---

* The eastern Roman Empire continued as the Byzantine Empire, centred around Constantinople, until its collapse in 1453.

Treverorum, founded by Emperor Augustus around 18 BCE. It grew into one of Rome's grandest colonies, the administrative centre of Belgian Gaul, with a colossal population numbering some 75,000, and a full range of classical amenities: amphitheatre, circus, baths and bridges. The Porta Negra was the grandest arch in the region (a hulk still stands) and the cathedral that stands in modern Trier calls itself the oldest church in Germany,* built on the stones of a Roman predecessor; this resonant place was where all the German kings were crowned for hundreds of years.

Charlemagne's empire went on to occupy more or less the exact terrain bounded by our rivers: the triangular Burgundian realm from Lyon to Mainz to Milan. But its heart lay in the Latin-German cities of the Rhine: Strasbourg, Speyer, Worms, Mainz, Cologne . . . these were the temples of early Christendom. The Rhineland had long been the outer boundary of Rome's ambitions, now it was the natural heart of its northern existence.

Strasbourg was its great bridge over the river, and its extravagant Gothic cathedral was begun in 1009. Known as the 'coffee pot', due to its high roof and spout-like single tower, it commands a panoramic view over a 20-mile stretch of the Rhine. The world's tallest building until 1874, when a church in Hamburg surpassed it, it has dominated the Alsatian skyline for 2,000 years. It is a treasure chest, too, from the stained-glass windows showing nine Holy Roman emperors to the astronomical clock, which blends earthly time with the revolving stars. The present model is a nineteenth-century replica of the original, but still, at noon, three kings march out and bow to the infant Jesus.

---

* It has some interesting modern distinctions – it was the birthplace of Karl Marx. But even today it is the ancient echoes that predominate.

Strasbourg Cathedral: throne room of the Gothic Rhine.

The cathedral at Speyer is in some ways even grander. Built in the eleventh century on the hill above the river, to protect it from flooding, it is the largest surviving Romanesque building in Europe, and the burial chamber for twelve German monarchs, four of them Holy Roman emperors.

The triumph of Speyer, though, lies in its inventive use of architecture as a form of storytelling. The enormous front entrance is deceptive: in fact it has only one small, gloomy door into the cathedral's interior. But since this is carefully positioned on the west side, visitors are forced to pass through this dark, narrow gateway and then walk towards the full, shining glory of the surprisingly small (and therefore surprisingly bright) east window facing the Rhine. On any given morning, this design allows visitors to experience a powerful symbolic journey from darkness to light.

The great cathedrals of Mainz and Worms are just as ancient – built in 1181 and 1230 respectively. And both have vivid places in Europe's past – Frederick Barbarossa launched the Third Crusade from Mainz, while Worms was the stage on which Martin Luther stood his ground when invited to retract his heresies. And though, in this context, the great Gothic cathedral of Cologne was a relative latecomer, it is just as resonant. The tallest twin-spire church in the world, it was begun in the thirteenth century primarily to provide a home for a holy relic Barbarossa found in Milan – the bones of the Magi or the three kings (the number is not known, but it has been supposedly three for so long that it feels almost like a fact) who travelled from the east to pay their respects to the newborn Jesus. Barbarossa had the relics sent from Milan to the Archbishop of Cologne by oxcart (the river passage was far too dangerous). And suddenly the Rhineland was a place of devout pilgrimage for Europe's faithful, and the medieval Church had a new spiritual home.

One of the odd things about Cologne's past is that because it was a free-trading city in the days of the Holy Roman Empire, its archbishops were not permitted to live in it – they were electors for the *state* of Cologne rather than the city, and resided in Bonn. Moreover, since it lay on the west bank of the Rhine it was often French.

None of this could disturb the power of its holy relics. The bones of the Magi are still in Cologne Cathedral today, housed in an extravagant golden sarcophagus behind tall railings on its high altar.

Similar tensions led to work on the cathedral being abandoned in the sixteenth century. In fact it was not fully completed until 1880, which meant that for several centuries it had a wooden crane up on its roof, like a nesting bird, waiting for the angels needed to complete it.

While the Holy Roman Empire began as a colony, and became a partnership, it was soon more: the ceremonial heart of Europe. Three of the seven German electors (the potentates charged to choose the Holy Roman emperor) were Rhineland clerics: the archbishops of Cologne, Worms and Trier – and a fourth, the Count Palatine, was from close by.* So although formally the emperor still had to be anointed by the pope, this was becoming a ceremonial gesture: it was clear from 994 on, when Otto was able to propel his cousin, the duke of Carinthia, to become Pope Gregory V (the first German head of the Catholic Church), that power had migrated north.

The Church's loftiest councils were held in Cologne and Worms. These also became the centres of crusading zeal and anti-Jewish fury. Mainz, meanwhile, hosted what has been called the greatest feast in medieval Europe, a Pentecostal event in which 200,000 people

---

* The other three represented Bohemia, Brandenburg and Saxony.

fought, drank and took part in cathedral ceremonies surrounding the knighting of Barbarossa's sons.

And so the Rhine did for Germany what the Rhone had done for France: it imported the two great products of Rome – classical civilisation and the Christian heritage that succeeded it. Both had deep and abiding effects.

20

ADD HOCK

Along with its sturdier effects – walls, bridges, roads, forts and theatres – Rome had also brought a range of new delicacies to the Rhine: melons, cherries, figs, apricots, chestnuts and peaches all thrived. But wherever the river bumped up against hills the Romans were quick – just as they had been in the Rhone Valley – to grow their favourite fruit: grapes.

Rome, Christianity, wine – the same story unfolded in this northern part of the empire, flooding down the Rhine to Charlemagne's Holy Roman domain. In fact it was from his palace at Ingelheim that Charlemagne noticed the speed with which snow was thawing on the south-facing slopes above Rudesheim. He had shown a similar feeling for vine cultivation in the Burgundy village of Corton, which is why that became home to the superlative white wine known as Corton-Charlemagne.

Soon there were as many as forty wine villages on the middle Rhine. And since the princes and bishops were happy to take their tithe in the form of wine (which they could sell) it rapidly became a commercial as well as a sacred product. By the fourteenth century there were sixty-two customs posts on this stretch of the river, all of them designed to dip their beaks in the lucrative trade.

Today's vineyards clad the Rhine-facing slopes all the way from Switzerland to Bonn – at first on the eastern side, in Baden, but on the western Pfalz escarpment too. A notable wine trail takes in the highlights of Rheinhessen and Nahe – the wines from Nierstein, Bacharach, and the walled garden of the Liebfrauenkirche in Worms. The last of these became a British favourite – *Liebfraumilch* – partly for its fruity sweetness and partly because of the poetry in its name ('Milk of Our Beloved Lady').

Further north the river swings into the famed *terroir* of Moselle, another steep valley with perfect conditions for grapes. But it is in the stretch of water between Mainz and Koblenz – the Rheingau – that the Rhine becomes a serious vineyard. And something more, because this is the gorge that smiles from the cover of a million cruise-ship brochures, the place where the Rhine ceases to be a freight super-highway and turns into a merry ribbon swishing between ruined castles and haughty cliffs.

And between those outcrops, on every inch of workable soil, stand the vines. Without the Rhine it would not be easy to grow grapes this far north, but, here as elsewhere, the river was a thermal control system and a mirror for winter sunshine. Where the river turned west, the right-hand bank tilted south to face the sun. It was too frosty a latitude for the Syrah grape, but for Riesling it was ideal.

That is how this became a distinguished wine terrain.

As always, the monasteries played a prominent role. The Cistercian Kloster Eberbach, founded near the Rhine in 1136, was one of the first to plant Riesling grapes, and by the sixteenth century its biggest wine barrels could hold 10,000 gallons apiece. The monks also planted red grapes from their Burgundian homeland, and gave birth to 200 sister foundations on the Rhine, most of which were wine producers too.

Rudesheim, from Bingen. A fairy realm of vineyards and spires.

The Benedictine abbey at Trier, meanwhile, had seventy-four vineyards to help it justify the ways of God to man. This was more than cottage gardening.

And there was another factor behind wine's success in this riverine landscape. Along with its basic gifts – soil, water and sunlight – the Rhine offered two extra ingredients: a distribution network and a ready supply of well-heeled customers. The cities of the Rhineland provided both, and because of the ease and convenience of transport down the Rhine, the wines from this region rapidly conquered England. Falstaff loved to marinade his tall stories in jugs of Rhenish, and thanks to the Hanoverian monarchy, anyone who was anyone in eighteenth-century London wanted to quaff German wine in the dining rooms of Mayfair and Pall Mall. When travellers started visiting the vineyards of the Rheingau, they stopped in the region's main port, Hochheim – opposite Mainz – and the wine became 'Hock'.*

Queen Victoria visited Hocheim with her German husband in 1845, inspiring one local winemaker to create a vintage in her honour. The result was 'Königin Victoriaberg', a Riesling still served on royal visits – Charles III may well have had a sip or two when, as Prince Charles, he attended a presidential dinner at the Schloss Bellevue in 1995, and it was also served to Prince William and Kate when they visited in 2017.†

Quite a number of ancient houses have survived into modern times – the signs to Schloss this and Schloss that on the road that

---

* Hochheim is actually just *off* the Rhine – its full name is Hochheim-am-Main, and it lies up the river that leads to Frankfurt. But it was the major port of the Rhineland's wine trade in the eighteenth century, which in English eyes was close enough to be Rhenish.

† The grower, Georg Michael Papmann, went so far as to build a Victorian Gothic folly in the heart of the vineyard. It has appeared on the label ever since.

runs by the river are a tangible reminder of the area's roots in earlier times. One of the best known is Schloss Johannisberg, on the hill east of Rüdesheimer – a rolling plateau that might be the slope on which Charlemagne watched the snow melt. It promotes itself as 'the world's first Riesling winery', and has indeed been bottling wine for 900 years. The twelfth-century monastery built by the Benedictines included a Romanesque basilica named after John the Baptist – hence Johannisberg.

After many vicissitudes – this was the stretch of the Rhine that suffered most grievously from the wars that smashed through these parts down the centuries, like summer thunderstorms – the estate was handed over to Austria's Prince von Metternich in 1816 as a reward for his diplomatic contribution to the war against Napoleon, and it remains in the Metternich family to this day.* The baroque palace on the crown of the hill houses an A-list concert hall, a church and a smart restaurant – on summer evenings you can hear the hum of conversation as diners clink glasses and enjoy the superb view down to the river. But in the folds of these luxurious surroundings you may still hear the whisper of olden days, and the heady power of the drink swirling in those goblets.

And far below, like a silver thread, gently marches the Rhine.

---

* Though he served Habsburg Vienna, he was very much a child of the Rhineland: born in a castle near Koblenz, and becoming a law student in Strasbourg.

21

DREAMING SPIRES

The foot of the Po Valley is a fetid zone which, like the Rhine and the Rhone, has long been a flood plain. But while the volatility of those rivers was down to their raw speed and power, in Italy the culprit was simple topography. This is marginal terrain; it is as much water as land, only a few feet above sea level and sometimes well below it. The serpentine river slithers through, turning the province into a netherland that for most of its history has festered with plague. There were terrible floods in 589, 1150, 1438, 1882, 1917 and 1926 (just to name the low points).

The river that runs through this drowned world is not a thing of beauty: it oozes through the farmland like a drain. The historian Jonathan Keates has called it 'the key to [Italy's] very existence as a nation', yet when brochures advertise the 'magic of the Rhine', the 'treasures of the Rhone' or the 'glories of the Danube', the Po is absent. Its bridges are not Renaissance jewels; there are no Lorelei cliffs or ruined castles. It is a sea of silos, pylons and warehouses. Even the 'Venice-by-Water' cruises end their journey at Mantua; for anyone wishing to continue towards Ravenna and Padua, 'transportation is by motor coach'.

The most obvious side-effect is that the cities of the region are unusually dispersed. Only Ferrara is actually 'on' the river, and even that is situated (like Milan) on a patch of raised ground just 30 feet above sea level and south of the main channel. The rest are sited much further off, in the wider Po Valley.

The cities themselves are extraordinary, however. Bologna is a forest of medieval towers marinaded in gastronomy. Ferrara, seat of the d'Este family, is studded with villas, palaces and the world's oldest wine bar, dating back to 1435.* Mantua, in its lakes, is the birthplace of opera and home to the likes of Virgil and Monteverdi. Padua has its Giottos, its basilica and its ancient botanical garden. Parma, sacked by Attila the Hun in 452, was reborn as a city of libraries, baptistries and Parmesan. Verona is home to Juliet's balcony, the painter Veronese and a gaping Roman arena. Vicenza was the drawing board for the architectural legacy of Andrea Palladio himself, the designer whose columns and pediments adorn London from Chiswick House to the Covent Garden piazza. We haven't even mentioned Cremona, where Stradivarius made his perfect violins. And of course the diadem in this crown, still gleaming on her wooden stilts out in the green lagoon: Venice.

Ironically, it was in part the unreliability of the river that encouraged these cities to develop and flourish as they did. The constant floods laid down a layer of rich soil perfect for the agricultural products that were the basis of the region's wealth. And by providing easy terrain for the trade that would further fill the ducal coffers, the river created conditions that were ideal for the commissioning of cultural wonders. All of these enviable strengths grew out of the super-fertile Po Valley.

---

* Jacob Burkhardt called it 'the first modern city in Europe'.

And then came the region's most spectacular fruit of all: the Renaissance. The Po Valley became an exhilarating centre of learning and scholarship, a cloistered yet vibrant domain in which all sorts of advances could germinate. The damp land was a seedbed not just for rice and cheese, but for ideas, too.

~

We tend to think of the Renaissance as primarily an aesthetic matter. But it was a scientific revolution, too, and this aspect of it burned especially brightly here because this was where Europe had created one of its most important institutions: the university. It was not uniquely European – the mosque at Fez, in Morocco, had been a centre of scholarship back in the ninth century, and the so-called 'Golden Age' of Islam, with its great capitals in Andalusia, Baghdad, Egypt and Timbuktu, was a renaissance in itself. But the modern idea of an independent seat of learning dedicated to the pursuit of secular knowledge was born in this hazy plain.

It is usually thought that creativity flares in certain places and at certain times (Athenian tragedy, Gothic architecture, Shakespearean theatre, Impressionist Paris, Hollywood cinema and so on) for financial reasons. Creative expression requires patronage; inspiration alone is not enough. But simple geography plays its part too. The scattered cities of the Po – the cosmopolitan mix of Gallic, Roman and Germanic culture, with recent notes from France and Austria – were able to absorb both the financial muscle of the Roman Church and the mercantile vitality of Florence. Eastern wisdom was trickling this way too, through Venice, so the whole region was alive with enticing new prospects and possibilities. The result was an outpouring of ideas that had no precedent.

But those ideas needed vessels in which to incubate. Europe's first university was founded in Bologna in 1080 by a group of students who hired a learned elder to lead and supervise their discussions. It soon counted four popes among its alumni, and it wasn't long before there were other such institutions in nearby Vicenza (1204), Pavia (1361) and Ferrara (1391).

Nowhere, however, did the flame burn as bright as in Padua. The college founded there in 1222 catered for the sons of Venice (daughters were not welcome then) and soon became *the* place for Aristotelian logic. It had a renowned Professor of Surgery, Andreas Vesalius, whose 1543 work on anatomy, *De Humani Corporis Fabrica* (*The Fabric of the Human Body*), became one of history's leading medical textbooks. His famous arrangement of human bones was donated to another university, on another river, where it became 'the Basel Skeleton'. It is still there.

One of his fellow professors was the Polish stargazer Nicolaus Copernicus. The son of a copper merchant in Kraków, he had studied law in Bologna and Rome before arriving in Padua in 1501 to theorise —heretically – that the sun, not Earth, stood at the centre of the universe (and imply, in passing, that perhaps *Rome* need not be taken to be the centre of the world, either).

His successor was Galileo Galilei. Born in Pisa, he too came to Padua as a professor – in 1592. He stayed until 1610, a period he called 'the best eighteen years of my life'. It might have been in Pisa that a swaying chandelier made him grasp the principle of the pendulum, but it was in Padua that he made the celestial observations that proved Copernicus right. In the short term this merely enraged the Church, and led to his being confined to house arrest for the remainder of his life. But it made him an immortal figure in the history of science.

Word travelled fast – and far. So when a young English student named William Harvey was preparing to graduate from Cambridge in 1597, he faced an interesting choice. He was invited to stay on at Gonville and Caius College, but Padua now had an even more prestigious school of medicine, complete with a botanical garden and purpose-built 'theatre' for anatomical investigations – the finest such facility in all Europe. Since opening in 1595 it had made Padua a pinnacle of science.

Harvey made his way to Padua in 1599, graduating as a doctor of medicine three years later. It was thus in the Po Valley that, studying the flow of blood in humans and frogs, he observed that it must 'circulate'. The notion that the human cardio-vascular system mimicked the cyclical motion of water wasn't *entirely* new. A hundred years earlier, in his Milan notebooks, Leonardo da Vinci had explicitly made the connection. 'While man has within him a pool of blood,' he wrote, 'so the body of the earth has its ocean, which also rises and falls every six hours with the breathing of the world.' He added that 'the cause which moves the water is like that which moves the humours in all the shapes of animated bodies'.

This insight also drove Leonardo's sense of the 'perfect' city, a carefully planned grid whose channels, locks, gates, pumps and valves did the work of a heart and veins. At the time this was a startling notion. But Harvey's experimental dissections forcefully demonstrated that it was more than just a theory. By examining in detail the way blood circulated through the human body, he changed medicine for ever.

Not many intellectual leaps have so starkly transformed the way people think – with one stroke of a scalpel he punctured the clinical rationale for bloodletting. And yet it was just one of many discoveries coming out of Padua at that time, emerging from the estuary of the Po, whose lively university culture was pumping new ways of thinking across the land.

And since nothing is as mobile as a new idea, it didn't take long for similar organisations to put down roots elsewhere. Avignon led the way on the Rhone in 1303, followed by Orange in 1365, Aix in 1409 and Valence in 1452. On the Rhine, Heidelberg came first, in 1386, to be followed by Cologne in 1388, Basel in 1460 and Mainz in 1477.

There was another transforming fruit of that early spread of knowledge: mapmaking. This budded first and most fruitfully in the Rhine delta, where the cartographer Gerardus Mercator trained at the University of Leuven, near Antwerp. Founded by Papal Bull in 1425, the college attracted students from France, Germany, England and elsewhere, and its reputation was growing – Erasmus himself studied there before going down the Rhine to Basel. It was also cultivating the habit of giving its students Latin names – Geert Kremer who bent over his Greek, theology and mathematics became Gerardus Mercator.

After a two-year stay in Antwerp (a seafaring capital where it was impossible not to dream of faraway horizons), it was back in Leuven that he started to produce his famous maps. But in 1569 it was in Duisburg, on the junction of the Rhine and the Ruhr (a significant trading post for timber, fur, grain and salt), that he produced the rectangular projection that flew around Europe's capitals and inspired a thousand successors.

His map's chief virtue was that it represented the world in a way that made it practical – creating the map on a cylinder rather than a sphere gave it a grid of straight lines, ideal for maritime navigation. It thus underpinned many of the later discoveries regarding latitude and longitude. But even more important was its psychological

impact: through a mixture of sheer volume and mathematical preci-
sion (Mercator produced hundreds of maps and charts of Europe
and the world) he changed the way Europe thought about itself simply by giving people the means to visualise it.*

Revolutionary ideas such as this, born in Europe's fledgling university sector, are further evidence that the scholarly retreats born
in the Po Valley really were the foundation stones of the Scientific
Revolution. Its pioneers are honoured to this day. When astronomers
mapped the surface of Mars in the 1970s, they followed Mercator's
principles of projection. The EU's Earth Observation Programme,
meanwhile, which oversees space projects, marine safety and climate
change, is named Copernicus – and its satellite navigation system
(a £10 billion project involving twenty-five orbiting craft) is named
after Galileo, who has also lent his name to a lunar crater, an airport (in Pisa), an unmanned spacecraft, a telescope (in the Canary
Islands) and any number of schools, colleges, ships, streets and urban
squares. Closer to home, the William Harvey Hospital in Ashford,
Kent – not far from the great man's birthplace in Folkestone – has
a 'Padua' ward in his honour.

The water cycle thus supported an intellectual cycle, which in
turn looped back into practical scientific knowledge. These ideas and
these new ways of thinking were relayed along Europe's major rivers
(still the continent's primary transport system) like the shower of
sparks from a firework, spreading the discoveries of the Renaissance
all across Europe.

---

* Mercator's house on Duisburg's Oberstrasse was shattered by Allied bombers in
the Second World War.

22

THE CONDUIT OF THE REFORMATION

If the Rhine could have only one emblematic character, it would probably be the man in whose shade Mercator grew, and whose course of life was influenced by the same geographical currents: Desiderius Erasmus. Posterity remembers him as Erasmus of Rotterdam – with justice, since that is where he was born (between 1466 and 1469). But he travelled widely in Italy and England before making his way to what was then the brand-new University of Basel, founded in 1460 some 450 miles upriver from his Dutch origins.

Rotterdam has an Erasmus Bridge, an Erasmus Medical Centre, an Erasmus Gymnasium and an Erasmus University, while his Basel house and gravestone are tourist highlights. But he also spent time in Cologne, Mainz, Strasbourg and Konstanz. His shadow falls over the entire length of the river, and in return it gave him the broadest possible horizons, and also helped spread his ideas throughout the continent. The convergence of new universities and new printing technologies, flowing together like two unstoppable rivers, encouraged some of the early intellectual cloudbursts that informed both the Reformation and the Renaissance.

Basel was the right place for such a man. Standing on the cosmopolitan corner of France, Germany and Switzerland, it was one of the first places to receive and absorb these new ideas.

Erasmus was also decisively helped by his friend Johann Froben (better known as Frobenius), who graduated from Basel University in the 1480s before becoming the city's foremost publisher. In 1513 he produced Erasmus's *Adages*, commissioning a cover illustration – an allegorical figure of humanity in a chariot (steered by Cicero and Virgil) – which expressed the central ideas of humanism.

The two men became so close that Erasmus actually lodged in Froben's house. And that is where, a few years later, in 1516, they worked on their groundbreaking Bible, using a newly discovered Greek manuscript to refresh and polish the old Latin 'Vulgate' version that had been used by the Catholic Church for a thousand years.*

Their work contained the shocking implication that the Bible was open to commentary and interpretation. Even without his notes, the mere fact of the twin translation hinted heretically that the Word of God might be provisional, imperfect, ambiguous – and could come in many voices. It suggested that scholarship might even be *superior* to faith, a higher court. It also allowed a wide readership to notice that at no point did the Bible have anything to say about popes and priests. Its very existence was an affront to some.

It was not quite the beginning of a new way of thinking. Protests against the corruption of the papacy were already widespread. But

---

* The fourth-century translation that St Jerome, born in what is now Slovenia, had produced from the original Hebrew manuscripts.

the Erasmus Bible put the relationship between God and Man (as humankind then called itself) on an ambiguous new footing. In a torrent of letters and tracts from his seat on the Rhine, Erasmus dramatically reinforced the way of thinking that would develop, very soon, into the Reformation.*

He was not the first person to put forward the basic tenets of humanism – these were already blowing on the breezes that swept out of the universities and up the rivers. Two hundred years earlier, on a different river (the Rhone), another original thinker, Petrarch, had been similarly inspired.

Born in Italy in 1304, Petrarch had grown up in papal Avignon, in the Comtat Venaissin. As the son of a lawyer retained by Clement V, he was a close enough observer of the papal scene to call it 'a sewer, where all the muck of the universe collects'. This was free-thinking of a new and dangerous kind, and although he would go on to become one of the world's most notable scholar-poets, his greatest find – a collection of Cicero's letters unearthed in Verona Cathedral – was one of the first and boldest fireworks not just of the Reformation (as a critique of the Church) but the humanist sensibility that begat the Renaissance.

And he had yet another claim to originality. When he clambered his way up Mont Ventoux, a vigorous day's walk east of Carpentras (and now a gruelling highlight of the Tour de France), his chief motive was 'the wish to see what so great an elevation had to offer'. He was duly rewarded with a glorious sight that was in those days a novelty – the view of Europe from a rarely attained height. No one at that time climbed mountains for the sheer pleasure of it – why

---

* He was not lazy. It is estimated that he wrote roughly forty letters per day.

would they? But from the crown of Ventoux, Petrarch could see the Alps, the wide Rhone Valley from Lyon to the sea, and the green hillsides of the Cévennes.

Yet in a letter written fourteen years after the event, in 1336, he recalled not the view, but the fact that at the top he had found only rocks and weeds, and he had occupied himself by reading the copy of St Augustine's *Confessions* that he always had about him for companionship.

'And men go about to wonder at the heights of the mountains,' he read, 'and the wide sweep of rivers ... but themselves they consider not.'

The words made him shiver. Far from enjoying an enhanced sense of worldly beauty, this high-altitude experience on Mont Ventoux was instructing him to shun the physical world, and seek the truth within.

This was one of the earliest and greatest insights of humanism. Wisdom lay within, not in the haughty teachings of the Church. The first stirrings of this radical new idea, which would have huge consequences both for the religious life of Europe (through the Reformation) and for its cultural sensibility (the art and architecture of the Renaissance), were first conceived of in an intense flash of perception on a mountaintop above the Rhone.

And now, in Erasmus's day, that insight was flowing down the Rhine as well – as fast as the river itself. By sixteenth-century standards, the Erasmus Bible was an unprecedented bestseller – new editions appeared in 1519, 1522, 1527 and 1536; soon there were 300,000 copies in circulation. It was the book that launched a thousand schisms.

It meant that for the second time (the first being the Gothic invasion of the Latin empire) the Germanic north was turning the

Roman world upside-down. The river that brought Erasmus to Basel took him away for a while when he travelled to London in 1509 to stay with his friend Thomas More. In fact, the idea for *In Praise of Folly* – a playful satire that he finished writing in a week after arriving in London – came to him during that long river voyage. Folly, he mused, as his boat carried him through Cologne and down to his old home in Rotterdam, was not the enemy of wisdom, but the soul of wit. 'No party', he wrote, in a snub to the prevailing theological mood, 'is any fun unless seasoned with folly.'

It is partly this errant career that has made him a symbol of the continent as a whole, and that is why he is immortalised in the EU student exchange known as the Erasmus Scheme. The title neatly doubles up as an acronym: the European Action Scheme for the Mobility of University Students. It has had a profound impact on European togetherness, and not just in an intellectual sense. Since its launch in 1987 it has produced (or so the EU likes to claim) more than a million Erasmus babies.

$\sim$

To understand what made all this possible we need to go downstream for a moment, and a few years back in time, because although the spread of radical ideas through Europe was partly to do with a shift in the intellectual weather, it was also down to the new technology available along the Rhine. And if Erasmus is synonymous with Rotterdam and Basel, then the Rhine cities of Strasbourg and Mainz are associated with an equally inventive genius: Johannes Gutenberg. As the pioneering father of European printing, he gave Erasmus and his followers the means to distribute their ideas and insights to the widest possible audience.

The documentary evidence is slight (inevitably there were no printed records) but Gutenberg seems to have been born in Mainz in or near the year 1400. The son of a merchant, he was well educated, but little of his childhood is known. He was raised in the old Roman city on the banks of the Rhine, opposite the Main (the route to Frankfurt), which in those days would have been full of barges ferrying wine to England.

It seems he spent a decade in Strasbourg, and it is fair to assume that he took his first steps towards developing a printing press there, because when he returned to Mainz in 1448 he immediately borrowed start-up funds from a man named Johann Fust, using his 'instruments' as collateral. His extraordinary new machine – a pair of boards that could be squeezed together by a vertical screw – was inspired by the wine presses he had seen and known since his youth. Two years later his print shop on the Humbrechthof (opened in partnership with Fust) was turning out pamphlets, poems and grammar guides by the barrow load.

This was where, in or around 1455, Gutenberg manufactured his famous forty-two-line Bible, the first book printed using movable type – a system by which individually crafted metal letters could be arranged into blocks of print. Of course it is now known that books along these lines had emerged in China and Korea many centuries earlier – the oldest printed work with a credible date of birth is a Buddhist woodblock from the ninth century. But Chinese, with its thousands of symbolic characters, was not well suited to movable type. Europe's concise and supple Latin alphabet was much more open to mechanisation.

Regardless, Gutenberg's 'invention' was revolutionary. His mechanical press could stamp out hundreds more pages in a day than the most industrious scribe could copy in a year. Moreover, Gutenberg's design left blocks of white space to allow for the addition of hand-coloured capitals, meaning that each copy came in its own individual trim. It was a giant step forward into the age of mass reproduction, and the consequences were overwhelming.

When you look at that historic first Bible in Mainz's Gutenberg Museum today – and there are two of them, sensitively displayed in a windowless chamber, air-conditioned like a bank vault – one thing stands out: its size. It is the dimensions of a small suitcase, weighing about two stone.* It was not, in other words, meant to be portable; on the contrary, it was designed to be chained to the lectern of a church or a monastery desk. And while it did not usher in the democratisation of Christianity in a single flash, it did not take long for its revolutionary potential to become clear. Word of the book, and then the book itself, fluttered along the Rhine like the flame on a fuse. It was ignited by a conflict between rivals for the bishopric of Mainz in 1460, which turned into a nasty war between the Elector Palatine and Baden, creating a diaspora of inventive craftsmen. Faced by the threat of occupation, Mainz's printers decided to try their luck elsewhere, and they took their tradecraft with them. Soon they were inflaming imaginations all over Europe.

Gutenberg himself took little part in that expansion. Only a year after the production of his Bible – an electrifying moment in Europe's history – he became entangled in a financial row with his partner. Fust wanted his loan repaid, and Gutenberg wasn't in a position (or willing) to do so. A 1456 lawsuit found in favour of Fust,

---

* The book's sheer weight saved its life in 1969, when a thief used a rope to climb down from the roof of Harvard's Widener Library in order to steal its Gutenberg Bible. He made off with both volumes, but they were so heavy he lost his grip on the rope on his way out, tumbled six floors and fractured his skull. The book was unharmed.

and Gutenberg had to hand over the entire shop. Fust teamed up with an apprentice named Peter Schöffer, and set about producing the world's *second* machine-printed book: the Mainz Psalter. This was the first to use colour – mostly red initials – and also the first to feature a publisher's brand or colophon. Gutenberg, the innovator, had omitted to leave any sort of stamp on his own achievement.

The partnership between Fust and Schöffer was a happier affair – Schöffer married Fust's daughter – but they soon had competition. Mainz became a rich warren of print shops, as did Strasbourg, Basel and Cologne. Before long, there were printing presses in Lyon, Paris, Nuremberg and Venice.

In 1463 Ulrich Han (from Ingolstadt, on the Bavarian Danube) produced the first printed book in Rome, indeed in all Italy: a woodblock-made prayerbook titled *Passione di Cristo*.* A press was operating in Basel in that same year, and soon, in 1467, two lay brothers, Arnold Pannartz (of Cologne) and Conrad Sweynheim (from Eltville, on the Rhine near Mainz), set up in a Benedictine monastery near Subiaco, east of Rome. There they published a Latin grammar along with works by Cicero, St Augustine, Caesar, Livy, Ovid, Pliny, Suetonius and Virgil. The recovery of classical thought did not happen overnight, but neither did it take long. From now on, timeless treasures were accessible to anyone who could read (and had the leisure needed to bend over books).

It was a world-changing technology with dazzling side-effects. One of Gutenberg's most notable innovations was a simplified new font designed to be more readable than the traditional Gothic script.

---

* When a surviving fragment came to light in Munich in 1927, it caused a sensation. It was bought by Princeton Library.

It was inspired by and called Roman, and did not just enter the visual vocabulary of European thought – it helped define and shape it.

It is well understood that the Renaissance emerged out of the classical works that the Church had not been anxious to see publicised. But it was not just the pressure of all that ancient wisdom that advanced European life; it was the machinery on which those rediscoveries could circulate. The humanist upheaval that inspired Europe's growing class of educated citizens to question both their relationship with God and their relationship with each other may have been initially prompted by the quest for rare books, but this alone would have come to nothing without the advances in engineering through which the new thoughts raced.

The cork was out of the bottle. Knowledge, ideas, art, music . . . sparkling ideas flowed round Europe like champagne. In 1470 the Rector of the Sorbonne, Joseph Heyntin, invited three Rhine printers to Paris. Valencia produced its first book in 1473, Antwerp in 1474, Lyon in 1476. And that same year a youngster from Kent (perhaps from Hadlow, near Tonbridge) went to Cologne to learn the art of printing from the masters. Gutenberg had died in 1468, but the Rhineland was still the centre of excellence.

That Englishman's name was William Caxton, and when he returned to Westminster he opened the shop that printed Chaucer's *Canterbury Tales* (in 1476), Malory's *Morte d'Arthur* (in 1485) and dozens of other works by Gower, Lydgate, Aesop, Cicero and Ovid – many of them in his own translations. It wasn't only printing that began on the Rhine; English literature can be said to have been born on its banks, too.

These early books were only the first glow of a starburst that brought civilisation out of its crypts and cloisters and into the drawing rooms not of the common man (mass literacy was not yet widespread enough to permit such talk) but of the merchant class. Before Gutenberg, there were 25,000–30,000 books in Europe. But by the end of the century there were an estimated 6 million, and by 1600 that had grown to 150 million. It was the best-aimed blow to established authority since the sack of Rome.

Naturally, Mainz has an imposing statue of its favourite son, and has also dedicated an annual festival, a *Platz* and a museum to Gutenberg's achievements.* But he is more than a local hero. In 1999 he was named (in a US poll) the most important European ever, and a year later was *Time* magazine's 'Man of the Millennium'. It is not an overstatement to call his printing press the most significant invention in Europe's history. Not for the first or last time, the Rhine was the epicentre of European life.

And it wasn't only books. Perhaps without intending to, Gutenberg also invented the news. Mechanised printing made it possible for people to keep up with the latest pamphlets, journals and notices. And his timing was good, because the world was full of good stories.

In 1492, for instance, a fiery streak appeared in the sky over Alsace, and caused a sensation when it thudded into a field near Ensisheim – on the French side of the Rhine north of Basel. It was a meteorite, but to the crowd that gathered around the smoking rock it seemed that God was throwing stones.† Fortunately, it had crash-landed close to the waiting presses of the Rhineland. Sebastian Brant, a Strasbourg-born poet now in Basel (and author of a satire, *The Ship of Fools*, a model for Erasmus's *Praise of Folly*) was quick

---

* After the Second World War, in honour of its extravagant history as a Roman capital, a medieval archbishopric, a cathedral of wine and the birthplace of printing, Mainz was awarded the honour of being officially twinned with . . . Watford.

† The 250-lb rock is displayed in Ensisheim's local history museum to this day.

In the beginning was the word. Gutenberg guards Mainz Cathedral.

to produce a news sheet featuring woodcut imagery of the incident. The meteor went on to appear in the *Nuremberg Chronicle* of 1493, and might even have inspired the falling star in the background of Albrecht Dürer's engraving *Melancholia* in 1514.

It was, literally, the first news flash. But not the last. A Genoese mariner had sailed across the Atlantic. The Moors of Islam had been thrown out of Spain (along with the Jews). John Cabot had reached New Found Land. Bartolomeo Dias and Vasco da Gama were pushing past the cape of Africa. Savonarola was on fire in Florence. Michelangelo was painting the Sistine Chapel. Spanish conquistadores were marching across Mexico. Leonardo da Vinci had moved to Paris. Erasmus had been spotted having dinner with Thomas More.

The world was growing more extraordinary every day.

In the Bible room of the Mainz museum we can see the design of the future's newspapers in the typographical layout of Gutenberg's first volume. Each page of his Bible included two tall blocks of hyphenated and justified type. They look exactly like modern newspaper columns.

And there was a further aspect to this advance. By setting in motion a revolution in reading, writing, learning and thinking (and questioning) Gutenberg was giving life to another new industry: paper. The monks who made the Lindisfarne Gospels in the eighth century needed 130 sheep to provide the skins that became their parchment – they could not honestly say that no animals were harmed in the book's creation. But since each of the 180 two-volume sets in that first edition amounted to 1,286 pages, Gutenberg required 230,000 sheets of paper. There was not enough paper in all Mainz for such an immense undertaking, which is why forty-five of those first 180 Bibles used old-fashioned vellum made from calf skin instead.

Paper-making was not brand new – there was a mill in Nuremberg in 1390 (the first north of the Alps), and by Gutenberg's time Strasbourg and Basel were producing paper too. But after this momentous stride forward things moved fast. And since mills needed power (to drive the hammers that mashed rag and wood chip into pulp) they needed to dip their wheels in water. The Rhine and its tributaries were perfect sources of energy, which means that the water rushing out of the Alps provided not just the pattern of settlement and communications on which the new ideas could run, but the physical energy needed to turn their engines.

~

In 1512 Pope Alexander VI, seeing the danger, excommunicated any printer who was acting without papal approval. But it was too late. The new publishing infrastructure was firmly in place on the day, in 1519, when Martin Luther pinned his famous Ninety-five Theses to the door of Wittenberg's church. He was reacting in part to the notorious sale of 'indulgences' by Vatican salesmen, who were touring Europe in search of funds for the latest restoration of St Peter's Church (by Bramante) – wealthy members of the flock were invited to buy their way out of their other penances in a scheme that in some ways resembled a modern bond issue. But thanks to Gutenberg's printing contraption his defiant message flew across Europe.

In truth it was not the pinning of his argument to the church door that launched the Reformation. It was written in Latin, the language of the elite; most of the passers-by were illiterate, and the authorities are unlikely to have tolerated such inflammatory material for long. The theses would almost certainly have been torn down within hours. Aware of this, Luther took the precaution of sending his grievances to the Archbishop of Mainz directly, in Gutenberg's home town.

The presses were waiting. Luther was able to read a printed copy of his own outburst within weeks. A year later, he wrote a pamphlet that did not simply denounce the papacy but rejected the idea that it held spiritual power over him or anyone else. The presses pounded. Word spread.

It didn't take long for the Church to feel the tremors, so it summoned Luther to the famous 1521 Diet of Worms. There, with the Rhine rocking past, Luther was given the opportunity to withdraw his words in front of the Holy Roman Emperor, Charles V.* Luther's obstinate refusal – 'Here I stand; I can do no other' – crackled through the presses of the Rhineland like a starting gun, sparking the upheaval that would tear Christendom apart. Its proposal that faith be a matter of private conscience meshed nicely with all the humanist ideas pouring off the printing presses.

Worms was no stranger to such convocations – its riverside location had long made it a proud imperial city. And while the Bishop's Palace where the meeting took place no longer exists (it was destroyed by Louis XIV's army in 1689), a plaque in the garden marks the approximate spot where Luther refused to budge. The monument features a bronze-splitting bolt of lightning, to symbolise the schism, and a pair of 'Luther's shoes' allows visitors to walk in his footsteps for a reflective moment or two.

A few hundred yards away, in the tree-fringed Lutherplatz, stands the Reformation's largest monument, the Luther *Denkmal* (or 'think piece'). Some 20,000 people gathered in 1868 to see it

---

* The fact that the emperor was newly crowned and inexperienced only adds drama to the scene. At the meeting to decide the fate of Europe, Charles V was only twenty-one.

unveiled: Luther on the central plinth, ringed by eleven other heroes of reform. Some are famous names – Jan Hus, Philip Melancthon, John Wycliffe – and there are also allegorical busts of cities such as Magdeburg, which suffered grievously in the religious wars that lay ahead (25,000 massacred), or nearby Speyer.

But the most striking feature of the memorial is the fact that so many of its figures are clutching books. Philip I of Hesse and Frederick III of Saxony brandish swords, and Hus a cross, but Luther himself, his friend Melancthon, the scholar Johann Reuchlin and the English cleric and Bible translator Wycliffe . . . all are armed with Gutenberg's new weapon.

The printed page. The most powerful siege engine of them all.

By 1525 close to half a million works by Luther were in circulation – and read aloud they would have reached a congregation many times that size. Luther knew this, and continued to write feverishly (his collected writings amount to fifty volumes). But in a 1522 pamphlet he acknowledged his debt to Gutenberg's machines by insisting: 'I did nothing; the Word did everything.' His German translation of Erasmus's Bible put the Word of God, which for centuries existed only in Latin (on pain of death) into the hands of ordinary people, allowing them to explore theological issues according to their own inclinations. Some 5,000 copies of this Bible were sold in the first two weeks, and the then-radical notion that faith – along with virtue and redemption – might derive from inner struggle rather than outward observance (the insight that came to Petrarch on Mont Ventoux) was from then on all but impossible to resist.

As Victor Hugo later put it in *The Hunchback of Notre Dame* (1831): 'Before printing, the Reformation would have been just a schism; printing made it a revolution.' Alas, he did not write his own admiring book about the Rhine (*Le Rhin*) until a decade later.

Perhaps that is why he did not add that without the river, and the unsettled, cosmopolitan fever with which it swirled between its Roman-Franco-German-Swiss-Dutch banks, there might well have been no printing presses in the first place.

And since the Rhine wound its way through Luther's career as well, from his show trial in Worms to his pamphleteering in Mainz, it really can claim to have been a central thread in the story of Europe's governing ideas. Gutenberg . . . Erasmus . . . Luther . . . Mercator: what a mighty set of tributaries. And they all splashed out of the Rhine.

23

THE RHINE GORGEOUS

When it leaves Bingen and turns north into the hills, the Rhine enters the picturesque gorge beloved of the river cruise supplements: a fairy-tale realm of ruined castles and abbeys. It is the narrowest part of the river between Basel and the sea. And as the Rhine curves between its cliffs, the valley takes on the menacing air of a Norwegian fjord – eyes narrowed and shoulders hunched.

Today it is a World Heritage site. The cruise boats haul a quarter of a million people a year round these bends, 60,000 of them British. And there is plenty for them to take in. As they putter between Bingen and Koblenz – a 50-mile trip – they pass twenty-five castles, two of which are fortified islands. Between them these hold a thousand stories. That one was razed during the Thirty Years War, the next one by Louis XIV. This one survived two sieges but was pulverised in the Second World War; that is where a corrupt bishop was eaten by mice; over there is where Queen Victoria spent a night on holiday with her German husband. Signs recall the high-water marks, from the 'great' flood of 1850 to the time the river froze in 1963.

Only one of these castles – Marksburg – has never been destroyed. The rest have been repeatedly ravaged. That is now their charm. One of the least damaged is the nineteenth-century neo-Gothic Drachenburg, near Königswinter. At various times this lavish folly has been a seminary, a Nazi college, a US military school, a refugee camp, a railway training centre and an amusement park. The only thing it has *not* been is a fort.

The rest have adopted many guises – palaces, prisons, museums, vineyards, armouries, hunting lodges, council offices, hotels, spa resorts, private residences and – these days – restaurants and photogenic models.

Perhaps the most spectacular is Ehrenbreitstein, Brentano's birth-place, the haughty fortress overlooking Koblenz. Grandly positioned above the Rhine–Moselle confluence, it has been in turn a Bronze Age camp, a Roman garrison, a medieval castle, an archbishop's pal-ace and a state treasury – sometimes French, sometimes Prussian, always intimidating. Today it is a museum, an archive and a youth hostel, with a cable car that hoists tourists up and over the Rhine for an ice cream and a city view that is still breathtaking, though the city below has been much bombed.

With so much heritage on display the Rhine has long been allur-ing to visitors. But historically it was also renowned as a place of extreme peril, both on the water and off.

~

The gorge came to prominence in the twelfth century when Frederick Barbarossa established an imperial toll on the Rhine near what is now Düsseldorf. The river was then the heart of his Holy Roman Empire – and he was supported by its cathedral cities. But this was an empire rooted in dynastic and religious allegiances rather than in territory. There was no such concept as the nation state. Even war was a private matter.

But the Rhineland was a wealthy prize, and though its princes owed their dues to the elected or anointed emperor, in practice they were highwaymen, and these castles were toll stations (just like Stockalper's tower on the Simplon). The barons would throw chains across the river to bring boats to a halt and lean on them for tariffs.

In theory, Barbarossa's customs post had a lofty purpose – to 'enlarge justice and bring peace to all'. But few of the 'robber barons' (and some robber bishops) had anything resembling a high-minded motive, and in practice the imperial court was content to tolerate them in return for a share of the proceeds. Recognising the value of an orderly river, it regulated them, ordering that they be spaced a certain distance apart. And in time the barons came to agree that an organised system was preferable to a free-for-all, and formed themselves (in 1254) into a Rhine League. Like all guilds, part of its purpose was to prevent newcomers from breaking in.

It is easy to see what a tempting target this narrow stretch must have offered to medieval bandits and brigands – they could watch boats approaching from afar, and spring on them when they rounded the bend. The fact that control of the waterway was so appealing to pirates prompted the League to invite Rudolf I, Count Habsburg (then based in Switzerland) to police it, on the grounds that only an outsider could perform such a task. He had a fearsome reputation – he had been excommunicated for burning down an abbey in Basel, and undertook a crusade for penance. He was made king of Germany in 1273, and thus was launched the monarchy that would become Europe's greatest dynasty. The Habsburgs went on to be emperors of Austro-Hungary and, thanks to their possessions in the Rhineland,

Holy Roman emperors. Not for the first time, Europe's history was being written along the Rhine.

～

The story took a more peaceful turn in the eighteenth century when Grand Tourists began to see this landscape as sublime. As well as the castles, the sweet, gabled villages on its banks – fruits of the ancient trade in wine and salt – made it even more atmospheric.

This aspect of the river's character received lyrical backing in 1775 when Goethe visited Lahneck and imagined its warrior past in a poem titled 'Geistes Gruß', or 'Ghost's greetings': 'High on the ancient tower / Stands the hero's noble spirit.' Just as England was rediscovering the world of Arthur and his knights through the works of Walter Scott, the Pre-Raphaelite painters, and Gothic Revival architects, so Germany was rummaging in the myth kitty of its past in search of spiritual guidance.

In 1800 the musician-author Clemens Brentano set his ballad 'The Lore Lay' on this same stretch. One hesitates to say that he 'must have' read the ground-breaking preface to Wordsworth and Coleridge's *Lyrical Ballads*, published that year – but he certainly echoed their feelings about liberty and nature. Brentano was born in Ehrenbreitstein, the castle at Koblenz – and his poem described a golden-haired maiden, jilted in love, climbing the cliff above the Rhine to lure sailors to their doom.

It was a classic Romantic motif.

> To Bacharach am Rhein
> Where lives a sorceress;
> So beautiful and fine
> She tore men's hearts.

Cliffs and castles: Pfalzgrafenstein, in the heart of the Rhine Gorge.

A passing bishop, struck by her intensity, instructs three knights to escort her to a monastery. But at the top of the precipice she imagines she can see her beloved passing by on the river below, and plunges to her death.

Only her song remains – a lovelorn siren to seduce others to their grave.

In 1824 the poet Heinrich Heine gave the story wings. His Lorelei is more than merely lovelorn. Indeed her opening lines (*'Ich weiss nicht, was soll am bedeuten / Dass ich so traurig bin'** – an almost exact echo of Antonio's world-weary confession at the beginning of Shakespeare's *The Merchant of Venice*: 'In truth I know not why I am so sad . .') – chime with the fashionable spirit of *ennui* or world-weariness that Goethe identified as *Weltschmerz*.† Friedrich Silcher set Heine to music; Liszt and Schumann did the same; and by the time Victorian tourist boats were filling the Rhine Gorge the Lorelei had become an established myth.

The Museum of the Rhine in Koblenz has several paintings of Lorelei's golden locks tumbling down to the river, as well as guidebooks, tea cups, wine glasses, matchboxes, beer mats and meerschaum pipes. And in 2011, when a 300-foot-long tanker, the *Waldhof*, loaded with sulphuric acid, crashed just below the Lorelei rock, blocking the water and killing two crew members, myth-making merged with conspiracy theory to make the legend even more powerful. Today's boats slow down and bob in the current to allow passengers to take Lorelei selfies.

---

* 'I do not know what it means / That I should feel so sorrowful.'

† Literally: 'world pain'. The feeling of existential dislocation has its roots in Dürer's engraving of the slump-shouldered angel in *Melancholia*, and was widely taken up by Romantic poets and novelists as a sign of refined, sensitive dismay.

~

The Rhine Gorge has attracted legions of devoted admirers. In 1806 Friedrich Schlegel (a scholar-poet who did pioneering work on Indo-European languages, worked as a diplomat for Metternich and achieved fame as the translator of Shakespeare*) marvelled at the sublime scenery, declaring that it made his heart pound to see 'traces of human courage in the ruins of nature, bold castles atop wild rocks – monuments of the time of human heroes'. It called to mind, he wrote, the 'higher heroic days'.

Byron felt the same about 'the castled crag of Drachenfels' he described in *Childe Harold's Pilgrimage* (1812), which narrated the adventures of a medieval knight errant who bore a passing resemblance to himself. Drachenfels means 'Dragon Mountain', and had been the site of a castle built to defend Cologne, wrecked by Sweden during the Thirty Years War and never repaired. This was just the kind of epic history that the Romantic imagination devoured.

In 1842 Victor Hugo looked beyond this vision and saw the Rhine as a political inspiration. It was 'the flood of the warrior and the thinker', and its destiny was to unite these two in one supreme current. The Rhine, he felt, was 'a profound stream . . . by whose murmurings Germany is bewildered in dreams'. It could stir the blood as well as the heart.

Turner's delicate watercolours rendered the gorge radiant enough to attract a stream of British visitors – the 'William Turner Route' invites us to stand in twenty-six spots where he took up his brush. And Mary Shelley, who took this route to Switzerland, imagined

---

* The 'Schlegel–Tieck' translation became first a staple and then a classic of German literature, to such an extent that it led some to insist that Shakespeare was in truth a German author who just happened to have been discovered by the English.

her protagonist Dr Frankenstein haunting its 'ruined castles' and 'tremendous precipices'.

Geography here was not just history, but melodrama. Edward Bulwer-Lytton, Secretary of State for the Colonies and a *belle-lettrist* famous for quipping that the pen was mightier than the sword, swore, in describing his own journey up the river from Rotterdam, that the Rhine's 'waves flow from the clouds' and that 'the name of this river is TIME'.

The Rhine does not really *have* waves in the oceanic sense, though we know what he means: the ships throw up heavy washes. But utterances of that sort gave the river a folk-mystical air that made it irresistible to the more romantic sort of patriot. In time that would evolve into the dark music of nationalism, with often terrible consequences.

24

WAR AND PEACE

It sometimes seems as though Europe has been locked in permanent war with itself. For thousands of years its monarchies, churches, tribes, nations and revolutionaries were enmeshed in conflict after conflict. Invariably, many of these tussles took place on the banks of the rivers; the cities that flourished along the rivers – with their bridges, cathedrals and markets – have been both sought after as prized treasures to capture and feared for their might.

The Rhone has been relatively tranquil. Hannibal skipped across it without a fight; and though there were skirmishes over the Comtat Venaissin between royal France and papal Rome, they were modest compared to the more extreme violence that was erupting elsewhere.

There were savage battles along the Ticino and the Po, particularly in the sixteenth century, as France, Switzerland and the Vatican fought for Milan and the mountain road. In 1513 they clashed at Novara, in the heart of the rice fields – a murderous hand-to-hand wrangle in which the Swiss killed 5,000 French troops. And two years later, when the French had their revenge at Marignano, their triumph was in part aquatic: they opened the sluice gates and released the Ticino. The river did the rest.

'I have vanquished those whom only Caesar vanquished,' proclaimed the victory medal struck by King Francis I of France.

We cannot do more than hint at the weight and fury of the battles that have convulsed this region; millions have died in the fighting, and in the plagues and famines that followed. In its way, this has done as much to shape the Europe we know today as the peaceful arts we prefer to dwell on. And it is the Rhine, the bloodstained moat between French and German Europe, the rift that marks the line where Rome stopped, that has seen the worst of the continent's many uproars.

None of these were so horrifying – or formative – as the Thirty Years War, the continent-spanning mincing machine that involved almost all the great powers of Europe, and which left the entire landmass both devastated and ready to be reborn. It lasted – as its name reminds us – for an unimaginable three decades. And since the armies involved were private militias, most of the riverside cities were ransacked and left to starve. It made the wars that preceded it look like minor disagreements.

In one way, it flowed directly out of the Reformation – it was an angry reaction to the theological uprising. As we have seen, Gutenberg's new presses acting as an accelerant to both the hopes and the ill-feelings that simmered in Europe's corridors of power. What began as a religious clash – the imperial envoys to Bohemia were thrown out of an upstairs window by furious Protestants in what became known as the 'defenestration of Prague'* – soon erupted into a squalid tangle of competing ambitions. Austria, France, Sweden, Prussia, Spain: most of Europe's powers were sucked

---

* In some accounts the envoys landed in a pile of dung, a humiliation that saved their lives. In other versions they were saved by the intervention of angels.

into the bloodbath. England managed to hold itself aloof, but only by holding its own civil war on similar religious-political lines.

The entire Rhineland learned to fear the approach of armies. The warring parties all came this way, and for the same old reasons: greed, politics, military opportunism. They were mostly mercenaries whose only reward was pillage, so even the richest duchies could not sustain them for long. In the name of godliness, Europe indulged in a frenzy as gruesome as anything since the days of the Goths and Vandals. Freiburg, Oppenheim, Bacharach, Kreuznach, Heidelberg, Frankenthal: all the river towns were brutalised. The larger cities were jewels in the Holy Roman crown, and their locations on the water made them obvious targets – Koblenz had a menacing citadel, a superb prize. Mannheim actually became the headquarters of Sweden's warrior king Gustavus Adolphus, who saw the river as both a bridging point and a defensive stockade.

The whole landscape was plundered. Farms were ruined, crops seized, livestock butchered; townships burned; castles shattered; women tortured as witches; children taken away to be cannon fodder.

When it ended, Germany's population had practically halved. At the start of the war, the city of Magdeburg had 20,000 citizens; fewer than a thousand survived.

~

Like a river exhausting itself in a set of rapids before reaching a calmer pool, the uproar eventually subsided in the Peace of Westphalia. The series of treaties signed in 1648 were as ambitious as they were wide-ranging, and laid down a series of conditions that were both reasonable and historic.

The treaties required a protracted effort by 109 different delegations, and, since some of them could not be in the same room, meetings had to be held in different cities – Cologne, on the Rhine, and then Münster and Osnabrück in the northern part of Holy Roman Germany. Since the negotiations took four years, these cities had to be protected as neutral zones. Then it was up to the embassies from Austria, France, Sweden, Prussia, Rome, Spain and the rest to iron out their priorities and red lines.

It could hardly have been more cumbersome. Yet it ended up being every bit as historic as the terrible conflict itself, enshrining new notions of statehood that would alter Europe's course for good. Not the least of its fruits was that by merging the 350 small Germanic states into 39 larger units, it set in motion the long-term process of German consolidation. It was the first step on the road to unification.

There were many other stipulations that would have a long-lasting impact on Europe. The Holy Roman Empire was disbanded, replacing the old imperial structure with a more modern concept of devolved sovereignty. The papacy was required to withdraw from northern Europe and Catholics, Lutherans and Calvinists were declared to be equal (a blunt repudiation of the past – for the first time in centuries, religious freedom could be held sacred). The Swiss Confederation was formally recognised and Spain's dominion over the so-called Spanish Netherlands (effectively the Rhine delta) came to an abrupt end. Alsace and Pomerania were awarded to France and Sweden, giving France its heart's desire – a border on the Rhine – and Sweden mastery of the Baltic. And many barriers to trade were removed.

There were scores of other pledges, many of them pointing the way to a new path for the continent, but there was also an important moment in our river story: many barriers to trade were removed.

Although the Peace of Westphalia was able to demand only 'a degree' of trade freedom, it meant that the navigable Rhine was at last open for business.

It does an injustice to the Peace to see it as no more than a set of simple or palpable outcomes. There were many false hopes and inconsistencies, and some of its stipulations did not hold. But it was a major step forward to recognise the Rhine as a mutual resource, not merely something to be fought over. As we have seen, the princedoms with access to the river had formed various fledgling unions in the past, but ten years after the Peace of Westphalia, they came together to form a new Rhine League, a formal union of some fifty German principalities designed to create a more consistent community in this ruined part of Europe. Ironically, it was created not by the German princes themselves but by Cardinal Mazarin, the successor to Richelieu in Louis XIV's French court. And *his* motive was to build a bulwark against Habsburg Austria. But even as a buffer zone it brought the electors of Mainz, Cologne, Bavaria, Trier and Münster together, allowing them to develop the reflexes needed to become their own nation in due course.

And so the bloodiest era in the Rhine's history came to a close. It was certainly not the end of the conflicts that raged over this river, but it was a decisive break with the warmongering that filled its past.

25

ART AND BOOTY

When Napoleon emerged from the rubble of the French Revolution and rose through the ranks to become a military leader, he was quick to make his presence felt. As the crowned heads of Europe gathered to teach regicidal France a lesson about the power of monarchs, he gave the British a bloody nose in Toulon, then put down a royalist uprising on the streets of Paris. In 1796 he was given command of the armies of Italy and invited to take on Austria's imperial force south of the Alps. All the old wars along the Ticino and the Po in Lombardy had left Habsburg Vienna in charge of the region; France fancied it could change all that, and Napoleon turned out to have an eye not just for Europe's strategic riches, but for its cultural treasures as well.

Dashing into Piedmont, he won a series of clashes against the kingdom of Sardinia, then forced the Austrians to retreat to the Alpine bank of the Po. They soon found they were not safe there either. Napoleon crossed the river at Piacenza and surprised them at Lodi.

That makes it sound easy. It was not. To achieve that crossing, a manoeuvre then thought to be slow, cumbersome and easy to resist, he detached a general to distract the Austrians near Turin, seized a ferry near Piacenza, planted a bridgehead on the north bank and used barges to take his 7,000-strong army across the immense stream.

To the Austrians it was as if he was in two places at once.

Lodi was the battle in which he first became known as the 'Little Corporal', thanks to the speed and decisiveness of this deployment. The very next day, a delegation from Milan gave him the keys to their city. But he did not stop there. He chased the Austrians east, beating them in a series of battles that culminated in a famous victory at Rivoli. Then he attacked Mantua, the fortified capital that commanded the Po delta, the approach to Venice, and the main road north to the Brenner Pass and Innsbruck.

Mantua had seen war before, in 1629, when a ripple of the Thirty Years War ('The War of the Mantuan Succession') led to its being seized twice by Spanish armies. But it had enjoyed a century of peace until Napoleon arrived. In the eight-month siege that followed, 16,000 Mantuans died from a mixture of starvation and disease.

He was soon the master of Venice as well. Yet still he did not pause, urging his soldiers on until they stood a hundred miles short of Vienna. There, without even consulting Paris, he struck a bargain that exchanged his newest trophy – Venice – for the historic Austrian territories in Alsace.

It was a remarkable achievement. In a single peace treaty he had realised the old French dream of extending its borders east to the Rhine and south to the Po. It won him renown in a France that was tired of revolutionary infighting and hungry for leadership. He was not yet thirty.

While he was in Egypt, seeking to undermine the British presence in the Mediterranean, a Russian force led by Marshal Suvorov (an ally of Habsburg Austria) managed to take Lombardy back, leaving Napoleon no option but to return. Like a second Hannibal, he hurried back over the Alps, pausing in Martigny to assemble the mule trains needed to drag his cannons over the Great St Bernard Pass.

Suvorov retreated to Switzerland, but a large Austrian army remained in Piedmont, threatening the south of France. Napoleon's army poured out of the mountains and swept down the Ticino. Within days he had captured Pavia and Piacenza, then crossed the Po again and caught the Austrians at Marengo. Risking a battle before his strength had arrived was on the reckless side of daring, but reinforcements appeared in time for him to win the day.

It was later said (truthfully) that Waterloo was a close-run thing. But Marengo was even closer. The first courier from the battlefield, before that last-gasp turnaround, actually spoke of a French defeat. All of which only added to Napoleon's reputation as a miracle-worker. Within months he was First Consul, with powers (thanks to the way the Revolution had swallowed up independent fiefdoms) greater than any king.

$\sim$

These Lombardy campaigns were of serious geopolitical importance. They gave France back a heady part of its ancient *patrimoine*, built a rampart against Austrian aggression, spread the republican vision, and provided a new source of wealth – 60 million francs were sent back to Paris in the first year. But they were an inflection point in art history, too.

Within days of his victory Napoleon was writing to Paris requesting 'a few reputable artists to take charge of the choice and transport of the fine things we shall think fit to send'. And he made it a condition of his treaties that occupied cities surrender their finest treasures, giving a crude legal gilding to his appropriations, which in truth was nothing more than theft. Northern Italy was raked of its masterpieces.

Paris greeted the arriving works with cheering crowds and a victory parade.* And the city had an ideal home for them: the Louvre. Originally a hunting lodge (at a time when the first arrondissement was haunted by wolves, or *loups*) and then a royal gallery, it had been co-opted by the Revolution and inaugurated as a people's palace on 10 August 1793. This was during the year of the Terror, when the so-called enemies of the Revolution were rounded up and imprisoned or led to the guillotine. It was ghoulish: people could stroll among the oil paintings even as the rattling carts – the legendary 'tumbrils' – carried the condemned through the Tuileries, taking in the artworks before watching the execution.

The Louvre already contained 537 great works, including several masterpieces by Leonardo da Vinci, but the stunning Grande Galerie, a 1,500-foot-long hall on the Seine, had room for more. In the name of French *gloire*, universal education, political prestige and military entitlement, hundreds of Renaissance treasures were scooped out of the palazzos along the Po, which had become a wonderful receptacle for art.

It is estimated that the first year alone yielded a trove of some 200 paintings and 100 sculptures, and the pace did not slacken. There were a dozen drawings by Leonardo, five Bruegels that had been wrestled south in an earlier convulsion, works by Cimabue, Ghirlandaio, Mantegna and many others. Milan lost 110 pieces; Parma and Piacenza were forced to hand over sixty. Raphael's drawings for the *School of Athens* were lifted away from the Vatican,

---

* As part of the celebration, a Montgolfier balloon hung in the air above the Tuileries, just like the balloon that supported the Paris Olympic flame in 2024.

which also lost one of its best loved sculptures, the *Laocoön*. The *Jewel of Vicenza*, a silver model of the city made by Andrea Palladio, was melted down. The *Bucentaur*, the gilded state barge of Venice, part of the city's Ascension Day ceremony, was stripped of its gold, smashed to pieces and burned.* The horses of St Mark's Cathedral, the rearing sculptures that had been Venice's figurehead for five centuries, were installed on top of the Arc de Triomphe instead.

Perhaps the most spectacular trophy was Veronese's *The Wedding Feast at Cana*. This wall-sized canvas (33 by 22 feet) had been made for the refectory wall of San Giorgio Maggiore, opposite the Doge's palace in Venice, where it had hung for 250 years. Napoleon's curators took it down, cut it into three pieces and sent it to Paris by boat. The voyage took months, but in due course *The Wedding Feast at Cana* became one of the Louvre's star exhibits – although by then the great gallery on the Seine was no longer called either the Louvre or its revolutionary replacement the Musée Central des Arts de la République: Vivant Denon, the archaeologist who went with Napoleon to Egypt, was made director in 1802, and obligingly renamed it the Musée Napoleon (a name it had to renounce in 1814).†

The colonnaded road that runs past its front entrance was also renamed, becoming the Rue de Rivoli in honour of the victory that had secured so much of this art. It was soon one of Paris's busiest thoroughfares. Crossing it was not to be taken lightly.

~

* This is the boat, dating back to 1311, that appears in the paintings of Canaletto.

† In 1812 Denon made his own hunting expedition to the Po Valley, parading the artworks he captured through the streets of Paris as if they were lions and elephants.

Much the same thing was unfolding in the Rhineland, where Napoleon's campaigns led to 200 Flemish paintings being hauled back to Paris, including masterpieces by Rembrandt, Rubens and Holbein.

Napoleon's motive here was to cement France's border on the river. The German-speaking lands to its west had been French for more than a hundred years, having been seized by the armies of Louis XIV in the seventeenth century and then by Revolutionary France. Now, the 1795 Peace of Basel made them a French *département* administered by the *Code Napoleon*. In keeping with the Revolution's secular vocabulary, Strasbourg's cathedral became known as the 'Temple of Reason'.

It was a familiar story. In its role as the border between France and Germany, the Rhine was again the place where the tension snapped. Napoleon had long wanted impregnable natural boundaries: rivers, seas and hills. The Rhine's destiny was to be France's front line in the east. But history has a way of unseating such hopes.

After the shocking defeat at Trafalgar in 1805, which established British naval supremacy so thoroughly that it put to an end all French hopes of an invasion, Napoleon moved instead to rebuild the association of Rhine principalities set up by Louis XIV and his Italian-born chief minister Cardinal Mazarin. The new Rhine Confederation, a league of sixteen states (under French control, of course) would once again be a buffer zone against Prussia and Austria. But it was also intended to be a guarantor of peace. The founding document (*Confederation of the Rhine and the Dissolution of the Holy Roman Empire – 1 August 1806*) declared the hope 'that the French armies which have crossed the Rhine have done so for the last time, and that the people of Germany will no longer witness, except in the annals of the past, the horrible pictures of disorder, devastation and slaughter which war invariably brings . . .' Two years later, in 1808, Napoleon also formed the Magistracy of the Rhine, one of whose duties was to administer the river in a way that would benefit all of its residents: princes, bishops, farmers, fishermen and *viticulteurs* alike.

At the time, the Rhine Confederation was effectively a colonial act designed to bring Germany under French control – but it had the interesting side-effect of proposing the idea of unification to a previously disparate scattering of small states, and it didn't take long for the German lands to unite against Napoleon. So when he limped back through the snow following his failed attempt to conquer Russia, he found himself facing a coherent enemy and was trounced at Leipzig (in 1813) by the Prussian Field Marshal Gebhard Leberecht von Blücher. Retreating over the Rhine, he was pursued by the ageing Prussian, who (inspired perhaps by his opponent's own famous crossing of the Po) made himself immortal by ordering his troops to build a pontoon bridge made of boats at Kaub, between Bingen and Koblenz, using the island castle of Pfalzgrafenstein as a halfway house.* On New Year's Eve 1813, he led 50,000 soldiers, 15,000 horses and 182 cannons over the water and into France.

A bronze statue marks the spot, showing Blücher in a military coat, pointing the way to Paris. The terrace of Pfalzgrafenstein also has a tomb-shaped monument, while the Blüchermuseum, housed in the apartment Blücher used, lets visitors explore what happened in colourful detail.

This was the first decisive defeat for Napoleon – the setback that condemned him to banishment in Elba. But it was the Battle of Waterloo two years later, in 1815, that finally brought an end to the Napoleonic Wars.

---

* The castle, Blücher's headquarters while the bridge was being built, is now a hotel.

Unlike the Peace of Westphalia, the Congress of Vienna that followed was almost as concerned with winding the clock back to pre-Napoleonic times as it was with charting a new future for Europe, but under the chairmanship of Austria's Prince Metternich it did redistribute territory in the hope of securing lasting stability. Among its many provisions, Swiss neutrality was enshrined, significant pieces of the French-occupied Rhineland were restored to Germany, and France's hopes of a border on the Rhine were dashed. More to the point: all that kidnapped art had to be returned.

26

ENTER TULLA

Earlier we saw how the Swiss Rhone went through a series of 'corrections', intended to make it easier for humans to live beside it. Much the same happened to the Rhine. At the time Napoleon fell, the river was a confused tangle of channels, meanders, loops, coils and dead ends – a wetland with loose shingle banks. It looped and lazed towards the sea, and its banks offered little protection against flooding. Its villages lived in dread of deluges – their god was a river god.

The most violent of its many spills is thought to have occurred in 1374. There is no known death toll but it is commonly thought to have been the worst such disaster in a thousand years, so it could have been deadlier even than the calamity of 1651, which drowned as many as 15,000 near Düsseldorf. Since then, the Rhine has become a different sort of river: straighter, stronger and more obedient. The floods of 2021, widely seen as the harbinger of a climate apocalypse, killed 190 – which is horrendous, but not quite comparable.

In certain obvious ways the alterations to the river's course have made it less natural – the impositions of modern industry have driven away much of its wildlife – but in other ways it is more useful: a heaving corridor for trade, freight, power generation, transport and tourism.

So thorough a change would not have been possible had that unkempt wilderness not been brought to heel.

~

Taming the Rhine took many decades. The various Rhine Leagues assembled to administer the water had not tried to change the river itself. On the contrary, they ruled through an ersatz tangle of jostling interests as boisterous and ungovernable as the current that raced past their battlements. It meant that the Rhine remained a lawless no-man's-land until more enlightened thinkers began to see that it could be organised in a more rational way. Industrial-era science and logic suggested that the river could be turned into a servant – or a friend.

In 1789, just before the fall of the Bastille, the grandly named General Jean Claude Eléonore Le Michaud d'Arçon – a cartographer with a keen interest in military fortifications – went so far as to suggest nudging the river into a completely fresh channel. Like Louis XIV, he argued that a rearranged Rhine could be navigable, and also create fresh land for agriculture. A 1,300-foot-wide moat would carry the water in a powerful sweep past Mannheim and up to Cologne. But his plan was shelved in the turmoil of the French Revolution, and Napoleon's own commission to manage improvements to the river was ended by the Battle of Waterloo.

Enter Johann Gottfried Tulla.

Born in Karlsruhe in 1770, he was a child of the Enlightenment *and* a son of the Rhine – a man both of his time and his place.

His father was a Dutch-born brewer who moved up the Rhine, and as an aspiring scientist Johann studied chemistry at the University of Freiburg, the city of swans. After training in Scandinavia and Switzerland, he became part of the team in 1807 that 'tamed the Linth' – the badly behaved river that drove the textile mills of Lake Zurich. It was his first taste of Swiss engineering, and it taught him what it took to subjugate an undisciplined stream.

Back in Karlsruhe he set up an engineering college and by 1812 was hard at work on a plan to direct the Rhine into 'a single bed with gentle curves' – or even, in places, 'a straight line'. Digging out a new trench was a laborious task, to be sure, but the most difficult leap of all – envisioning such a thing – had now been accomplished.

Work began on Tulla's 'rectification' in 1817. The central idea was to increase the capacity of this stretch of the river by creating two beds: the new one, 650 feet wide and built with solid walls, like a canal, would run alongside the existing route. To reduce the risk of floods the most abrupt bends would be straightened, making the river faster as well as safer. More than 2,000 islands – some of them little more than gravel mounds, but still a menace to shipping – would be removed.

The scale of the project is still visible today: aerial views show the piercing line of the new route slicing through the old loops and curves. Tulla actually shortened the stretch of the Rhine between Basel and Bonn by more than 50 miles, a bold alteration that created as many problems as it solved: in certain places it improved the flow *too* sharply, making the Rhine too fast for sailing boats – barges, sometimes lashed into long and unwieldy trains, had to be hauled upriver by gangs of horses or mules. Fortunately for shipping agents, the age of steam was at hand, allowing boats to chuff up even the strongest stream unaided.

A river fit for industry: Tulla's straightened Rhine near Mannheim.

Tulla's work had other consequences. By improving the stability of the Rhineland, he unbalanced the river elsewhere, dragging water out of Lake Constance and endangering the low or 'nether' lands to the north. This raised awkward questions that required, and would soon call forth, a much more active approach to engineering and maintenance. His work thus created the conditions that encouraged – or demanded – a greater degree of co-operation between all the countries through which his river ran, reminding Europe that the Rhine truly was a joint venture.

The initiative that 'changed Switzerland' now found itself playing on a larger stage. No matter how sharp the differences between the bordering nations, the management of the river had to be seen as a common interest. It meant monitoring the level, the flow, the rate of sedimentation and erosion, and everything else. It meant working together to reduce pollution. And it meant agreeing standards on navigation, water use, safety, docking, towpath quality and a good deal more. The river was too precious to allow minor issues of national sovereignty to get in the way.

The Central Commission for the Navigation of the Rhine (CCNR) had already been founded in 1815. It grew out of an 1804 treaty imposed by Napoleon to organise the collection of tolls, and was based in Mainz and Mannheim. At first it was an agreement involving France, the Netherlands and several German states, but in 1919 Belgium, Britain, Switzerland and Italy joined as well. In 1920 it moved into the Renaissance-style Palais du Rhin in Strasbourg (near the Opéra du Rhin) and was given added responsibilities. It was an early league of nations, and its common ground was the river.

It didn't take long for further pacts to fall into place: a historic trade deal concerning the free flow of goods along the Rhine was struck in Mainz in 1831. It was the first of many. As the world's oldest international organisation (the Danube did not have a joint body until 1855, after the Crimean War, and the Compagnie Nationale du Rhone was not formed until 1933) it is theoretically possible to credit the CCNR as inventing the very concept of such cross-border agreements.

The idea of mutual advantage spread rapidly, not least because the memory of Napoleon's Rhine Confederacy was still making the German states shiver. Prussia was working hard to create its own free-trade area to address the fact that a commercial journey across Germany at that time – from Hamburg to Munich, say, or Cologne to Berlin – involved dozens of inspections and tolls. This daunting barrier to progress was swept away in 1834 when the forty-odd German states, plus Norway, Luxembourg and Sweden, agreed to remove the tariffs and turn Tulla's improved Rhine into a superhighway. The river thus gave rise to a clear belief that the key to wealth creation and national unity was the removal of trade barriers – a belief that would motivate European policy hugely in the coming years.

In 1840 the Grand Duchy of Baden signed an accord with France, and in 1892 Austro-Hungary and Switzerland agreed to regulate Lake Constance. France was granted permits for four power stations of its own on the Rhine. A 1969 German–French treaty even agreed to provide safe passage for fish. Tulla's locks and barrages had divided the river into two streams, with room for both shipping and turbines; this was the first occasion on which wildlife was included in the negotiations.

There were political consequences to all of this. Something like a Prussian-led empire was beginning to emerge out of the German lands in the east. It can hardly be said that these changes pointed the way to a peaceful future. The Rhine remained a diplomatic battleground even when no blood was being spilled – and there were still plenty of terrible conflicts to come. But the concept of the river as a shared resource was starting to present Europe with a glimmer of an idea regarding how, in theory at least, it might find a less bloody way forward.

That is why Tulla's achievement is so estimable: by turning the wild 19-mile-wide Rhine Valley into a waterway secure enough to support modern industry, he was advancing the idea of European togetherness a century and a half before Konrad Adenauer and Jean Monnet, the German and French architects of the European Economic Community built out of the ruins of the Second World War, sought to make it real.* The immediate effect may have been to make the Rhine an even more abrasive dividing line between French and Prussian pride. But it was at least understood now that the river smudged national boundaries. The Rhine belonged to six countries, but also to none of them.

~

Tulla's mightiest legacy, though, was the transformation of the Rhine into a key industrial resource. He died in 1828, but work on his rectification continued until 1871. And though it took a long time for it to become effective – in 1868 almost the entire Rhine Valley was submerged by floods – from then on the river became an altogether more reliable ally.

---

* Konrad Adenauer was the chancellor of post-war Germany, but before the war had been a long-time mayor of Cologne. He grew up on Tulla's Rhine, and had been mentored by the river all his life. Perhaps that is why it came naturally to him to make Bonn the capital of the new Federal Republic – it was just a few miles from his home, and he had been to university there.

This had consequences even Tulla might not have envisaged. Millions of acres of once boggy ground became fertile, and as the Industrial Revolution gathered pace, new harbours and quays opened up the Rhine to heavy goods traffic. The commercial reflexes of the whole region changed for ever. In due course the Rhine Valley would become the central spine of the so-called 'Blue Banana', a clumsy term chosen by economists to describe the curving industrial corridor that runs down the Rhine from Lancashire to Lombardy (it is more formally, and usefully, known as the Liverpool–Milan Axis).

To a greater extent than even he could have envisioned, Tulla's excavations extended the sea lanes into the heart of Europe. Switzerland, until then a rural hideaway that struck early travellers as an impoverished (if delightful) backwater – gained access to the trading power of the oceans. It would take one more step for the Rhine to become fully navigable – the completion of the Grand Canal d'Alsace, a 30-mile channel that ran alongside the river north of Strasbourg, opening up a shipping lane to Basel. Construction started in 1932, near Mulhouse, and when the canal was completed in 1959, it allowed 30,000 boats a year (container vessels, tankers, barges and everything else) to plough their way to Switzerland, and made the Rhine the continent's most important commercial artery. Some 300 million tonnes of freight now churns its way between the North Sea and the Alps.

Inland cities such as Karlsruhe, Mannheim and Strasbourg became major ports. The last of these, Strasbourg, 450 miles from the sea, had long styled itself the *carrefour*, or crossroads, of Europe; now that wish came true. Its enormous Rhine harbour, spread over 60 miles of waterfront, today handles more than 30 billion tonnes of freight every year. The cruise ship terminal alone looks after 800,000 annual passengers.

But it comes second to Duisburg in the list of Rhine ports by volume. As the service station for the iron and coal yards of the Ruhr, and now an important notch on China's ambitious 'belt and road' across Europe (trains stream into Duisburg from China every day), it is unmatched. It is an industrial powerhouse, with a long history as a manufacturing centre (textiles, tobacco and other commodities), and since 1891 it has been the home of Thyssen Steel, which on its own is a city-sized metropolis of blast furnaces and cooling towers.

Tulla's own birthplace, Karlsruhe, was for a while the southernmost port that could be reached without using a lock, transforming it (like Mannheim) from an aristocratic retreat into an industrial power, but it was superseded when the Canal d'Alsace gave the cities closer to the Alps equal access to the river's riches. And that was not the only such initiative. In 1833 workers completed digging the trench of an even more resonant waterway: the Rhone–Rhine canal. Running between Dijon and Mulhouse (where it meets the Canal d'Alsace) it linked two rivers that had long been emblems of rivalry. When the sluice gates were opened and the waters of France and Germany began to mingle and make up, a historic alliance was being conjured into being. In theory it was now possible to float from Rotterdam to Marseille, from the Baltic to the Mediterranean. Three subsequent Franco-German convulsions made traffic on the waterway difficult, but from then on the Rhine was always a powerful symbol of what could be achieved.

~

Tulla is mostly forgotten now. Any tourists wandering through the cemetery of Montmartre, in Paris, are likely to be after bigger game.

Armed with maps of the mortuary stones, they march along the green alleyways in search of Berlioz, Degas, the brothers Goncourt, Heine, Stendhal, Truffaut, Zola and other celebrities. But somewhere between Alexandre Dumas and Jeanne Moreau stands a quiet column dedicated to a 'Badois' (from Baden) named 'Jean Godefroy Tulla'.

It takes a moment to realise that this is Tulla's name in French. The rain-streaked plaque identifies him as the *ingénieur* who fixed the Rhine, and includes engraved images of his achievement – a book of drawings and a representation of his new channel cutting through the river's loops.

He is here because, although he was not French, he had studied in Paris in 1801. And it was Napoleonic France that took the initiative in wanting to improve the Rhine – effectively appointing him to lead the project. The bulk of his work was done after Napoleon's time, but France did not forget, and in 1827 he was awarded the *Légion d'honneur*. That inspired him to return to the city of his student years before his death.

His tombstone is an unvisited tribute to the idea of better together.

Tulla's role in the transformation of the river, however, is honoured in the *Tullaturm*, a watchtower in the Schlossplatz at Breisach, north of Basel. This old river town, with its steep hill and thirteenth-century cathedral, faces an island in the Rhine – making it a bridge as well as a fortress. But the most evocative memorial to Tulla is a single stone in a clump of trees near Karlsruhe, a mile-long walk from the nearest road. There are no signs to point the way, and though it had been raining on the morning I made the mini-pilgrimage to see it, it didn't look as though anyone had been there for a while – there were no footprints in the mud. The grass beneath the trees was a nettle patch, and though a plaque honoured Tulla as 'the man who tamed the wild Rhine', it didn't feel cared for by anyone – there were no flowers in the untended grass.

But perhaps the best memorial is the river itself. It marches past the ceremonial stone like a regiment on parade, buckles glinting, through banks of red poppies that nod in the breeze.

27

THE FACTORY FLOOR

A river, as we have seen, can be many things: a tap, a fountain, a fishing ground, an escape route, a border, a moat, a library of myths and legends, a power station, a place to live, a playground, a tourist attraction . . .

But it can also be something else: a rubbish tip. And Tulla's new Rhine, reconfigured as a corridor for heavy industry, brought this to the fore.

In the era before modern transport there could have been several feet of horse dung in the streets, and until the improvements in plumbing and sewage treatment that we now take for granted came along, all urban leftovers were simply dumped in the nearest ditch. And it was not only human refuse: animal carcasses, rotting fish and every other sort of waste was jettisoned in this way – perfect only for malarial mosquitoes and plague-ridden rats.

And the larger the metropolis, the greater the mess. Corrected, the Rhine was better able to flush away this refuse, but until it was finished, the maze of stagnant inlets and sluggish bends could still be overpowered by the sheer quantity of human dumping. That made cities like Cologne famously foul-smelling.

In 1834 Samuel Taylor Coleridge poked fun at it in a much quoted verse:

> The River Rhine, it is well known
> Doth wash the City of Cologne.
> But tell me, nymphs, what power divine
> Shall henceforth wash the River Rhine.

It was, he added, a city of 'two and seventy stenches, all well defined'. In fact, it had a Roman-era sewage system, a network of stone tunnels to carry refuse out of sight to the river.* But that had long since fallen into disuse.

Ironically, Cologne was already famous in another way: for perfume. We cannot say that the foul-smelling city actually attracted perfumiers, but it certainly made their products more necessary. In 1709 an Italian craftsman named Giovanni Maria Farina had left Piedmont, crossed the Alps and sailed down the Rhine to Cologne. There he changed his name to Johann and created an exhilarating new substance – an aromatic oil suspended in alcohol, with citrus notes designed to echo his Mediterranean heritage: 'My fragrance,' he wrote, 'is reminiscent of a spring morning after the rain: of oranges, lemons, grapefruit, bergamot and blossoms.' He named the new perfume Eau de Cologne, after its birthplace, and it was an instant success.

But it wasn't enough to disperse the unpleasant aromas wafting through the city. And as heavy industry and new machinery began

---

* In the Second World War these Roman drains were used as air-raid shelters. Today they are a tourist attraction.

to pollute both the air and the water, the problem was about to get worse.

~

After Tulla, things had changed fast. Mannheim welcomed its first steamship in 1825, and three years later was a free port – quite a snub to the Rhine's old role as a robber-baron tollbooth. By the end of the century Mannheim had four fully equipped docks. A drawing-room city of pianos and Biedermeier cabinets had rapidly become a manufacturing centre.

It was here that Karl Drais, a forestry official and part-time inventor, started work on his own revolutionary idea. In 1813, at the height of the struggle against Napoleon, he proposed a *Fahrmachine*, operated by a treadmill, that he hoped might replace horse-drawn wagons (and be especially useful in wartime, when horses were in short supply).

A few decades later the same idea was exploited by another local man, Heinrich Lanz, who developed this primitive concept into what we now know as the tractor. There were already steam-driven engines in Victorian fields, but this more mobile version, with huge wheels and a high platform, could cope with the most rugged terrain, and through the addition of a tow bar could deploy all kinds of tools. A century later, in 1956, the Lanz company was bought by the American tractor giant John Deere, and in 2023 its 2 millionth vehicle rolled off the production line.

It was the beginning of a tradition. Soon, Mannheim also became the city where, in 1885, Carl Benz developed his famous 'horseless carriage' – the motor car. The first model, a spindly three-wheeler, did indeed look like a pony and trap, with a wheel where the pony should have been and an engine under the seat. It had a top speed of just 10 mph and was none too reliable – in one of its first public shows it crashed into a wall. But Benz persisted, and the permit he received in 1888 allowing him to drive around Mannheim was in effect the world's first driving licence.

His cars were commercially available right away, and by the end of the century he was the world's leading automobile manufacturer (the first car in Britain was a Benz). Following the death of his partner, Gottlieb Daimler, he developed a new model with a designer whose daughter was named Mercedes – an ideal name for their motorised work-in-progress.*

Benz's wife Martha made an equally memorable contribution to this story by completing the world's first long-distance drive, steering her husband's *Patent Motorwagen* (apparently without his knowledge) to visit her mother in Pforzheim, 65 miles away. She packed her two sons in beside her and set off. On the way she noticed that the wooden blocks her husband had designed to slow down the gas-driven carriage made a truly excruciating noise, so she stopped, asked a workman to wrap them in leather, and invented brake pads. Later, when she pulled up outside an apothecary near Heidelberg to buy the solvent that was then required to power the engine, she was creating the world's first petrol station. A metal sculpture outside the shop recalls the moment.

The journey took all day, but it was historic. Martha is honoured not as Carl Benz's wife, but as motoring's first and most significant test pilot.

---

* After the First World War, Daimler became a British company and dropped all associations with the name Benz.

~

An even more dramatic transformation was under way on the opposite side of the river. The bog on the western bank had been handed to the kingdom of Bavaria after the Battle of Waterloo. It looked unpromising but the old fortress of the Rheinschanze had a very fine harbour, with plenty of room to expand. It had the potential to be much more than a customs post and was ideally located in the middle of the Rhine Valley.

Nothing much was done with it, however, until 1843, when King Ludwig I – the 'mad' Ludwig who built Neuschwanstein, a fairy-tale folly in the Alpine forest* – decided to develop it. As was his way, he named his creation on the Rhine after himself: Ludwigshafen.

While the Bavarian castle was a knights-and-damsels fantasy that looked only to the past, Ludwigshafen was a city of the future. In 1844 it was a fishing hamlet with a population numbering barely a thousand, but by 1849 the railway had arrived, linking the area to the Saar coalfield and detonating an industrial boom. When Ludwig's successor, King Maximilian II, released Ludwigshafen from its feudal dues in 1852 and recognised it as politically free, it became something completely new: a city entirely given over to manufacturing and trade.

The pioneer was a Mannheim businessman named Friedrich Engelhorn. A one-time goldsmith and the owner of a chemicals company, his bottled gas was greatly in demand for lighting the city's cafes, offices and streets; and the tar he obtained as a by-product (the gas came from coal) was proving valuable as a dye. After trying and failing to find premises for a larger factory in Mannheim itself,

---

* The spire-clad, faux-medieval inspiration for Walt Disney's 'magic' castle.

he bought land in Ludwigshafen. His company was named Badische Anilin und Sodafabrik – or BASF.

There are plenty of grim notes in its subsequent history. As Germany boosted its armament production in the build-up to the First World War, the company began to make munitions, explosives being close cousins of fertilisers – it was BASF that produced the mustard gas that spread so much terror in the mud-soaked trenches of Ypres and the Somme.

> Gas! GAS! Quick boys! – An ecstasy of fumbling
> Fitting the clumsy helmets just in time,
> But someone still was yelling out and stumbling
> And flound'ring like a man in fire or lime.*

Retaliation was inevitable, and in May 1915 Ludwigshafen became the first German city to suffer aerial bombardment when French planes raided the BASF works on the Rhine, killing a dozen citizens and causing serious damage.

Peacetime was dangerous, too. In 1921 a chemical explosion killed 500, injured 2,000 more and flattened half the factory. Even today, it is not rare to hear whispers in Mannheim that some of the firm's executives prefer to live at some distance from their own laboratories, for safety reasons as well as comfort (Heidelberg, Worms and Speyer being significantly more salubrious).

By the time war resumed in 1939, Ludwigshafen was one of Europe's noisiest industrial zones, with a population of 144,000, and BASF had grown into a global conglomerate: IG Farben. Three of its scientists won Nobel Prizes in the 1930s. At that time it was

---

* Wilfred Owen, 'Dulce et Decorum Est', *Poems* (London: Chatto & Windus, 1920)

denounced by Nazi leaders as a Jewish-led monstrosity, although in the war that followed it was ultra-loyal. It became a conductor of medical experiments in concentration camps and producer of the lethal poison – Zyklon B, a derivative of cyanide – used in the gas chambers. That (and its oil depot) made Ludwigshafen a target once more, and 10,000 planes flew down the Rhine to drop 50,000 high-explosive bombs on it. By 1945 there was very little left.*

Yet the power of its geographical location meant that even this near annihilation could not prevent Ludwigshafen from rising again. IG Farben was reborn under its old name. While nothing can hide its fearful history, today, thanks to the enormous Port of Ludwigshafen (a vast complex connected by a 10-mile-long railway) it is again the world's biggest chemical company, occupying its own 6-mile strip of Rhine quayside, with a dozen vessels coming and going every day. It employs 100,000 people in eighty countries, producing chemicals (solvents, acids, resins, glues), agricultural fertilisers, pesticides, plastics for automobiles, oil derivatives, films, packaging materials, and all sorts of biotech. It burns more gas than the whole of Switzerland.

The company even sells wine. The BASF Wine Cellar was launched in 1901 as an accessory to its core businesses, and is now a full-service merchant, stocking 2,000 brands from around the world and boasting that it can provide 'the right wine for any occasion'. Bottles signed by Steffi Graf and Helmut Kohl decorate the Ludwigshafen store.

---

* Many IG Farben executives swore, after the war, that they had no knowledge of these horrors. Zyklon B, it was said, was sold to labour camps in order to fumigate lice-ridden clothes. Not surprisingly, this has been much contradicted by historians and survivors for whom the true purpose of the product was 'an open secret'. In the Nuremberg trials, thirteen company directors were found guilty of war crimes.

A modern shopping mall, the Rheingalerie, has been built on the ruins of the old Rheinschanze, facing Mannheim. But at night, when its gas flares light up the sky, and steam rises from a thousand vents, Ludwigshafen still feels like a thunder god.

It was a child of its time – but also a child of its place. Had it not been for Tulla, none of this could have happened. This is why the town has installed a monument to the old hydrologist, a rust-coloured memorial fountain in a market square shaped like a bend in the river – and not just any bend: the 'Altrip corner' south of Mannheim, a very sharp bend in the river that was a famous nautical hazard until Tulla made it passable.

~

The fleet of cruise ships, freighters and barges hammering up and down the Rhine seems endless, and makes the river seem a living, writhing thing.* But there are few other signs of life. This is not a river for swimming in or pottering about on – you rarely see a sail, a canoe, a rowing boat, a paddle board or a jet ski. It is too fast for water sports. It is unusual even to see anyone sunbathing or reading a book on its banks, and the entire urban landscape is not designed to face the water.

It remains a feature of the Rhine that its cities do not overlook it; they are set back to avoid overspills, and few people actually live on its banks: you are more likely to see the chimneys of a power station or a business park than anyone licking an ice cream. There is the swimming dash through Basel, but after that, from Strasbourg to

---

* The long, slim barges, specially designed for this river, often have a car lashed on the stern, so the crew can explore the places where they have to stop for the night.

Rhineward-bound. A barge near Heidelberg heads for the sea.

Cologne, the Rhine is not a river for pleasure. It is too big, and too serious. For British visitors it is a marvel – a stream of liquid power, not at all like the quaint old Thames, easing through Berkshire with its bumbling launches, skiffs, punts and inflatable animals. The Rhine has weightier matters in mind.

But there is something else. The reason no one is fishing is simple: there are no fish. To a greater extent than Tulla could have imagined, his 'improvements' came at a price. Cologne may not have smelled pleasant in the good old days, but industrialisation exposed the Rhineland to a new level of pollution. Some of this was caused by agricultural run-off (fertilisers and pesticides); some was generated by the dumping of waste products; some was a simple by-product of fuel-powered shipping.

All Tulla had achieved, one glum commentator said, was to make a world safe for sugar beet. The fish fled; geese found quieter nests. According to Anne Schulte-Wülwer, Deputy Head of the International Commission for the Protection of the Rhine, in 1850 the river was home to 280,000 tonnes of salmon; but by 1950 there were none left. The Salmon Commission founded in 1885 to look after the river's fish stocks had nothing to protect and was wound up.

Not all the creatures banished by Tulla and his successors were mourned. No one missed the typhus and dysentery microbes that had flourished in the old marshland, or the mosquitoes that had made malaria a fact of life. And the man himself had been sensitive to the idea of conservation in principle. His 1822 pamphlet, *The Rectification from Basel to Mannheim, with a Justification for the Necessity of Regulating the River*, stressed the importance of a considerate approach: 'In all parts of the world,' he wrote, 'the lushest lands are to be found on the banks of rivers.' He knew as well as anyone that these needed maintenance: 'All matters and enter-prises that affect the climate and the fertility of the soil must be in proper balance.'

But he saw his work as enhancing this balance, not diminishing it. In his mind it was *nature* that was unbalanced. Freedom for the water meant disaster for the people who lived on its margins.

The ensuing lack of life in the rivers was only the start of the unforeseen consequences. From the middle of the nineteenth cen-tury on, the mines, mills, factories, wharves, power plants, chimney stacks and expanding suburbs – not to mention the run-off of agri-cultural chemicals – turned the Rhine, the grand old river of myth and legend, into a drain.

There was some recognition of this danger from the begin-ning, but environmental awareness was not a priority until after the Second World War. In the immediate aftermath, the Rhine was technically in the French zone, but there was a major American and British presence in Wiesbaden, so there was some openness to the idea of international co-operation. And the Netherlands, whose drinking water came from the Rhine basin, was quick to complain about the levels of potash and other poisons in its river.

That led, in 1950, to the formation of a new body, the International Commission for the Protection of the Rhine – a joint venture between Germany, France, the Netherlands, Luxembourg and Switzerland. In its early years it did little more than conduct surveys and write reports. There was a Federal Water Act in Germany in 1957, but it had no legal teeth until 1963, when the Bern Convention gave it the power to liaise directly with water companies and insist on certain standards.

This was the year following the publication of Rachel Carson's *cri de coeur* about the chemical despoliation of Earth, *Silent Spring,* a groundbreaking work dedicated to Albert Schweitzer, whose stark prediction that humankind 'will end by destroying Earth' helped it capture the public imagination. It sold 2 million copies and put the subject at the centre of political conversation for the first time. Doris Lessing hailed Carson as 'the originator of ecological concern'.

In 1966 another wake-up call arrived in the form of a Beluga whale, which swam 250 miles up the river to Duisburg, once the most polluted place in Europe. Christened 'Moby Dick' by the papers, it was pictured swimming past the chimneys of Germany's industrial heartland, avoiding the nets and tranquilliser darts deployed to capture it. Near Bonn it poked its head up, as if to get its bearings, while a NATO press conference was going on; concerned politicians removed their headphones and hurried out to take a look. The whale left the Rhine after a few weeks, but not before people had noticed that it was developing a rash on its white skin due to the pollution in the water. The world started to sit up and take notice.

Three years later, in 1969, a major spillage of the insecticide Endosulfan poisoned the Rhine at Sankt Goar, near the Lorelei rock. It killed millions of fish, and the bodies floated all the way to the Netherlands. Swimming and drinking were banned, and it triggered a hailstorm of bad publicity. The Rhine was denounced as a 'sick river' or 'the sewer of Western Europe'.

In 1970 the Federal Water Act was updated, and two years later came the first Conference of Rhine Ministers in Bonn. By now the planting of the European Community's major institutions in Strasbourg had made it keenly aware of the issue, and in 1976 it too joined the Commission for the Protection of the Rhine. Under Dutch pressure, new conditions were laid down for the mines in Alsace, which were putting potassium into the water. But the Rhine's woes weren't over yet.

In 1986 fire ripped through a warehouse full of chemicals in Basel. The blaze was controlled, and no one died, but the emergency hoses washed tonnes of hot chemicals into the Rhine. The river turned a livid shade of red all the way down to the Lorelei rock.

A few years later there was a 20-mile oil slick running past Düsseldorf towards the Netherlands. That led to a tougher regulatory regime, with new levels of monitoring. BASF started taking water samples for analysis every six minutes, every hour, every day, and there were increased penalties for dumping waste.

Slowly the efforts of the previous decades started to take effect. There were sightings of eel and trout, pike and perch. By 1996 the Rhine was declared to be Europe's cleanest river – fresh enough to provide drinking water for 20 million people. And the countries through which it passed remained committed to keeping it that way: the 1999 Convention on the Protection of the Rhine was signed by Belgium, the Netherlands, France, Germany, Luxembourg and Switzerland, with plenty of pledges about sustainable development.

However, with 10 per cent of the world's chemical industries based on the Rhine, it remains an accident in waiting and the complicated dance between environmental disasters and heightened regulations shows no signs of slowing down. In 2022, a report commissioned by the ICPR (published in German, French and Dutch) listed twenty major polluting events, most of them spills from the industrial plants along the river: an oil leak at the Rheinau power plant (ironically, a clean-energy provider), a foul-up at a chemicals factory in Leverkusen, an accident at the Krupp Mannesmann steel works.

The world has also been forced to notice a new threat: microplastics. Defined as particles less than 5 millimetres wide (smaller than peppercorns), they are an ingrained part of everything in which the Rhine specialises: packaging, bearings, textiles, electronics, cosmetics, automobiles. North Rhine-Westphalia alone has a thousand plastics companies, meaning that the river is thick with microgranules that are constantly being whisked into the current by ships' propellers. One 2015 academic study found the Rhine to be one of the world's worst offenders in this respect. Scientists took thirty-one samples between Basel and Rotterdam, collecting on average a million particles per square kilometre. Between Basel and Mainz there were 'only' 202,000, but higher concentrations existed downstream of the big cities (Strasbourg, Cologne, Leverkusen, Duisburg), and in the coal and steel empire of the Rhine-Ruhr region the level rose to a choking 2.3 million. The authors estimated that the river was spewing a 'daily load of 191 million plastic particles' into the North Sea.

Pollution remains a major issue on this industrial stretch of the Rhine. It will take something extraordinary to nurse it back to health.

28

WATER MUSIC

The Rhine Gorge attracted poets and dreamers. But it had an even more powerful influence on perhaps the greatest of Germany's arts: music.

It started long ago. In the early years of the twelfth century (the date is disputed, but it was between 1104 and 1112) the daughter of a well-born German family in the Rhineland west of Mainz was handed to the Benedictine monastery at Disibodenberg as an 'oblate' (an 'offered person') – a gift to God, not quite a fully fledged nun. In theory she would be free to leave at puberty if cloistered life disagreed with her.

It did not. She had been experiencing religious visions since the age of three, so a solitary life devoted to God did not alarm her. But in time (thanks to the Benedictine Rule, which insisted on a wholesome diet of serious reading) she became learned in many fields – botany, medicine and music as well as theology.* More important, she started to record the times the Holy Spirit visited her in dreams. Her name was Hildegard.

---

* Later she would deny that she had any scholarly leanings or skill, a modest claim that was not exactly borne out by the voluminous and erudite range of her works.

She was well-liked: her fellow nuns elected her Abbess in 1136. And in 1141 she was encouraged (by God, in a vision) to share her thoughts with others. 'Infirm human,' he commanded, 'say and write what you see and hear!' Word began to spread about this saintly lady on the Rhine. In 1148 the pope sent an embassy to evaluate her; reassured that she was not a heretic, he gave her his blessing.*

That didn't hurt. Retreating from the world made her a celebrity.

The women's house at Disibodenberg was at that time a junior wing of the male sanctuary, but Hildegard pushed to create an independent abbey of her own at Rupertsberg, facing Bingen on the Rhine. This would allow her to live a simpler and godlier life. When her superiors were slow to support the idea, she took to her bed, claiming paralysis by holy vision.

It worked. In 1150 she took twenty nuns to Rupertsberg, and when her book of visions was copied and distributed (no printing presses for her) she achieved fresh acclaim as a philosopher. To see her as a proto-feminist is to invoke a modern term that had little meaning in medieval Germany, where equal rights came a long way second to demons and heresies. But she was vehemently contradicting the norm. By now she was one of the most famous *people* in Europe. She corresponded with four popes and as many monarchs, and even walked to Ingelheim (a few miles up the Rhine towards Mainz) to share her visions with Frederick Barbarossa. Bernard of Clairvaux, founder of the Cistercian order, stopped to see her on one of his Rhine trips from Cologne to Frankfurt.

---

* It has been suggested that the visions were caused by migraine headaches, but this was patently too pedestrian an explanation for the twelfth-century Rhineland.

Her riverside location was crucial in that it placed her and her message on an important meeting of the ways, allowing her ideas to flow round Europe in a way they could not have had she stayed up in the hills.

But she saved her most fervent emotions for the music she wrote to accompany the abbey's roster of prayers. In this, as in all things, she insisted that she was only the mouthpiece of the Holy Spirit.

The music was what later came to be known as Gregorian chant – a single unaccompanied vocal line. And while it may be fanciful to imagine that it was directly inspired by the river, there is certainly something in the motion and rhythm of the Rhine – a steady, flowing pulse that never stops – that does suggest the music's onward march. Hildegard turned the gurgle of the mighty Rhine into a soaring expression of holy yearnings.

In 1228 the pope set in motion the procedure that would confirm her as a saint. But then she was forgotten. The humanists of the Reformation honoured her, but Rupertsberg was ravaged in the Thirty Years War, like most of the historic enclaves on this stretch of the river, and she faded from view. It fell to the twentieth century to reclaim her not just as a musical pathfinder but as one of history's greatest forgotten women. She was eventually canonised in 2012, nearly a thousand years late.

It says something about the way the world used to think that in the *Oxford Companion to Music* (1956) she is not even mentioned. But today her work is regularly performed, and she is a fixture at the top end of the classical music playlist – the best-known and most frequently recorded medieval musician of either sex: number one in Classic FM's chart of 'greatest women composers', and fourth in its list of all-time greats (behind, in ascending order, Beethoven, Mozart and Bach).

It is tempting to imagine that music was always her primary claim to fame. In fact it is the other way around: her work probably survived *because* of her prestige as a philosopher. Other composers matched her output; none enjoyed the renown needed to keep their work alive.

Today, the whole of Bingen is a monument to her memory – there are statues, schools, hospitals, churches and street names. The museum of her life is housed in an electricity-generating station on the river, behind a statue of Victor Hugo pointing at the view. On the hill at Eibingen, her abbey has been restored as a place of pilgrimage for tourists, with frescoes in the chapel showing scenes from her life, and a locally grown wine – Abtei St Hildegard Spätburgunder – stacked high in the gift shop.

~

The musical tradition she did so much to create came to full maturity in the nineteenth century, when a formidable set of composers emerged to form what amounted almost to a fraternity on the banks of the Rhine.

The first was Beethoven. Born in Bonn, in 1770, he grew up within walking distance of the Rhine – on a road named the Rheingasse, which ran down to the river. He would have seen it every day, and – to an extent made poignant by the fate that awaited him – he would have *heard* it. The sound of flowing water was an integral part of his sonic landscape, filling his ears with its lapping and slapping, tickling and trickling, now a gurgle, now a roar, plashing and splashing and whispering along in a tempo punctuated by fish, swans and boats. Before industry polluted the air with its engines, it was full of river-made sounds – the creak of mill wheels, the slap of boats, the ripple of flags and click of feet on paving. Beethoven's 'Pastoral' Symphony depicted the music of a country brook, but it is hard not to see it as the driving pulse in many other compositions.*

More to the point, the piece *was* a child of the musical and cathedral tradition nurtured on the Rhine by a thousand years of history. Over a century later, when Bonn became one of the capitals of the fledgling European Coal and Steel Community (thanks to the promptings of another local hero, Konrad Adenauer), it was Beethoven's Ninth Symphony that provided the dazzling theme – the world's first great supranational anthem – for the whole continent.

One of Beethoven's keenest admirers was the Hungarian virtuoso Franz Liszt. In the summer of 1841, Liszt visited an island in the Rhine – at Nonnenwerth, near Remagen – with Marie and their children.† They stayed in a convent that had been 'secularised' by Napoleon in 1802 and was now a country inn. Legend had it that Nonnenwerth was the resting place of Roland of Roncevaux, the knight-hero of the eleventh-century *Chanson de Roland*, who fought a rearguard battle for Charlemagne and retired to the Rhine to die. Liszt wrote an 'elegy' in his honour, *Die Zelle in Nonnenwerth*, in which Roland bewails his lost love, and also a song, 'Im Rhein, im schönen Strome' ('In the Rhine, in the beautiful stream').

The family's arrival caused quite a stir. Boats circled the island in search of a glimpse, and the Cologne Symphony Orchestra disembarked and staged a concert. On his thirtieth birthday Liszt planted a sycamore in the hotel garden, and came back in 1842 and 1843 to

---

* Obviously the music is not actually a description or representation of the river (though it does ripple like a current in the breeze); rather, it is an attempt to capture the 'awakening of happy feelings' aroused by the rural scene.

† One of whom, Cosima, would cement the musical link by marrying Wagner.

see how it was doing. It is still there as a visitor attraction ('the Liszt tree') today.

The next figure to take up the baton of Rhine music was Robert Schumann. Though born in Saxony, he studied law at Heidelberg and fell in love both with Romantic literature (Cervantes and Walter Scott) and with the river. In 1850 he became director of music in Düsseldorf (with his wife Clara and their seven children), wrote several pieces about the Rhine, met Heinrich Heine and turned the poet's work into a song-cycle.

His Third Symphony was 'The Rhine' or 'Rhenish', and took the river as its main character. The second movement, 'Morning on the Rhine', mimicked the wave-like ripples of Beethoven's pastoral river, while the fourth was meant to suggest a 'solemn ceremony' in Cologne Cathedral.*

In the end Schumann took his feelings about the flowing water to a melodramatic extreme (in 1854) by leaving his flat wearing a dressing gown, marching down to the bridge and jumping into the water. He did not drown – he was fished out by a passing boatman – but his wits never recovered, and he spent his last two years in an asylum a few miles up the river, near Beethoven's birthplace, before dying at the age of forty-six.

The composer most closely associated with the Rhine, however, is one who saw it with the eyes of an outsider. Returning to Germany from Paris in 1842, Richard Wagner wrote: 'For the first time I saw the Rhine – with hot tears in my eyes.' The sight provoked a patriotic spasm that led him to swear 'eternal fidelity to my German fatherland'. He would go on to compose an epic opera on this theme – and

---

* Both of the movements' explanatory titles were removed before publication so as not to let the music seem too directly descriptive or merely illustrative.

the poem he took as his guide, the *Nibelungenlied*, is one of the great legends of the Rhine.

Dating back to the twelfth century, it had been rediscovered in 1755 by a Swiss scholar-physician, Jacob Hermann Obereit, who came upon the forgotten manuscript in the library of the Hohenems Palace in Vorarlberg, on the Rhine near Lake Constance. The poem narrated the downfall of the Burgundian empire by telling the interlinked adventures of Siegfried the dragon-slayer, Gunther the Nibelung King, Gutrune his sister (Siegfried's wife), Brünnhilde the warrior-queen, Alberich the evil demon of the silver mines (Snow White and her seven companions owe something to this legend) and the 'golden hoard' of treasure salvaged from the Rhine and forged into an all-powerful ring.

Just as Switzerland's William Tell did not truly emerge until he gripped the nineteenth-century imagination of Schiller; and just as the scattered tales of King Arthur and Robin Hood were turned into English literature only centuries after their fabled deeds, so the *Nibelungenlied* did not truly become a central German legend until it rose from its grave in the nineteenth century, when national sentiments were coming to the fore. That is when it became Germany's archetypal fable – its *Iliad*.*

In Wagner's hands it became its *Odyssey* as well – the saga of a homecoming and the recovery of Germany's origins in a swirl of gods, giants, myths, heroes, flames and sacrifice. In time it would also give rise to Tolkien's *Lord of the Rings*, a story forged from the same magical elements (dragon, gold, ring, war, magic, destiny and revenge).

---

* It is sometimes forgotten that one of King Arthur's first adventures involved sailing to Brittany and using his own enchanted sword, Excalibur, to slay a dragon.

A sacred river, trembling with the memory of heroes, giants and gods, all wrangling beneath a blood-red sky – these were the ingredients that seized Wagner's myth-making imagination: the theft of the Nibelung gold from the Rhine and the heroic endeavours of Siegfried were an ideal frame for his larger purpose, which was to ignite a political revolution by remaking opera in a way that could forge Germany into an unconquerable host. In his eyes the Rhine was more than a waterway; by reviving this grand myth, Wagner made the river synonymous with German heroism.

Nation-building, set to music – this was something new: a solemn exercise in redemption, apotheosis and the death of gods. And three decades later, when *Das Rheingold* made its first appearance at Bayreuth in the autumn of 1869, just seven years after Bismarck vowed to unify Germany with blood and iron, it lit the flame of unification. *Die Walküre* followed in June 1870. And a few weeks later, in July 1870, Bismarck's Prussian army stormed across the Rhine to invade France.

~

As we have seen, France had long seen the Rhine primarily as a bulwark against Prussian aggression. But in Wagner's day it was the other way round. The aggressive campaigns of Louis XIV and Napoleon had not been forgotten, so as far as Germany was concerned it was *France* that was to be feared. Wagner took the Romantic depictions of the river poets, mixed them with a medieval epic, and created a banner of national pride.

The Ring is a work of sufficient majesty to resist any one simple interpretation. But Wagner was born in Leipzig in 1813, the place where (and the year in which) Prussia humbled Napoleon on his retreat from Moscow. And when France and Prussia squared up again in the 'Rhine crisis' of 1840, Wagner happened to be exiled in Paris at the time. He was not writing allegory, but *was* concerned to rouse German passions, and he used the myth-bespattered landscape of the Rhine to do it.

And if anyone wanted to see this tale of good and evil (pure gold stored in the river being filched by greedy dwarves) as a critique of the way the modern world was being corrupted by schemers and gold diggers – well, people were free to read into it what they wanted.

As it happens, the story of *The Ring* may be allegorical in a different way, since the medieval poem was itself loosely based on actual events. There *was* a Battle of Worms, in 436 CE, and in outline it bears an interesting resemblance to the story of the *Nibelungenlied*. In that fifth-century clash of arms, the Burgundian warrior king Gunther made a valiant, Siegfried-like attempt to push a gigantic army of Romans and Huns, united in a wicked pact against him, out of his homeland.

At times Siegfried has been seen as a fictional avatar of the heroic German leader Arminius, who massacred Rome's legions in the Teutoburg Forest after the death of Drusus, so *The Ring* is by no means an exact replica. But the other legend is still much in evidence in Worms today. The road that soars over the river towards the cathedral uses the Nibelungen Bridge. There's a statue of Siegfried outside the shopping mall, and a fountain of the Nibelungen people stands round the corner from the Luther *Denkmal*. Below the neo-Gothic tower on the city wall there's a Nibelungen Museum, and at the water's edge there is a bronze of Hagen, flinging the legendary golden hoard into the great river, where in Wagner's retelling it fell into the hands of the Rhinemaidens.

The Nibelungen Bridge at Worms: where Luther stood his
ground and the Rhinemaidens sang Wagnerian songs.

The Rhine flows through the centre of all these events. The whole fire-and-water pageantry of *The Ring* opens with the great river rising, and ends with it bursting its banks, submerging the funeral pyre of Siegfried and Brünnhilde as it reclaims the all-powerful ring.

Diehard Wagnerians (and no composer has such fervent disciples) do not enjoy discussing the notorious issue of Wagner's later role in the emotional life of Nazi Germany. They dislike the idea that he might have been one of its founding fathers, preferring to regard him as a genius co-opted by an evil cause. They are divided, in particular, by how much responsibility the music should carry for Nazi antisemitism.

But one has to bend quite a long way backwards *not* to see the story of *The Ring* – the seizing of the Rhine's gold by the dwarf Alberich, rising from the depths to snatch his prize – as an anti-semitic parable. The work does not specifically identify Alberich as Jewish, but it does not need to. As the greedy, malignant outsider poisoning the pure spirit of Germany, filching its gold and thwarting its forest-born heroes, he was, and is, a blatant and well-understood anti-Jewish caricature.

Nor is there any escaping the fact that the music struck the deepest of chords with Hitler himself, to the extent that his strident world view does appear to have been shaped by Wagnerian dreams. This is a large and controversial field of research; suffice it to say here that when Hitler met Britain's Neville Chamberlain, at the Rhine spa resort of Bad Godesburg, near Bonn, in 1938 – in theory to decide the fate of Czechoslovakia – Wagner's death mask lay on the coffee table between them; indeed, it was the only object there. Wagner was Hitler's theme music. The torchlit rallies at Nuremberg swayed to the overture from *Rienzi*, and in 1945 the radio station relaying news of his death played Siegfried's death march.

Wagner's Rhine thus ran through the Führer's life and his last days. National redemption through blood and suffering? It is not hard to see why he liked it.

~

The contested territory west of the Rhine, for so long a testing ground for French and German claims, still trembles with the long history of these clashing allegiances. Hochfelden, Reichshoffen and Oberschaeffolsheim are all in France. Betschdorf, Haspelschiedt, Langensoultzbach: each has a French flag hanging proudly from a pole in front of the town hall.

No wonder the architects of European unity chose Strasbourg as their residence – they knew they were stepping not only in the foot-prints of Frederick Barbarossa and Charlemagne, but on the bones of half a million soldiers. The glass castles overlooking the water in Strasbourg's political *quartier* – European Parliament, Council of Ministers and European Court of Human Rights – have their foun-dations in a blood-soaked past.

One of the most alarming moments in its already tense history came in the late eighteenth century, when the crowned heads of Prussia and Austria were gathering to suppress the fledgling French Revolution. An assault from the east seemed imminent, and France had yet to unearth its military genius – Napoleon was still in Corsica. The soldiers waiting to defend the bridges over the Rhine were anx-ious and fearful.

This provided another moment of musical inspiration. In Strasbourg's Place Broglie (a few yards from the opera house and the *mairie*) a stone tablet on the wall of a hotel recalls the moment in 1792 when Claude Joseph Rouget de Lisle, a junior French officer stationed there to defend the Rhine, composed a marching song to stiffen morale.

His composition was called 'Chant de Guerre pour l'Armée du Rhin', and he performed it in the house of Strasbourg's mayor, Baron Philippe de Dietrich – on Place Broglie, where the plaque now sits. It went down so well it became the subject of a famous painting, with de Lisle in full voice, arms flung out to impress the seated mayor. The mayor would be guillotined a year later (like most revolutions, France's was quick to devour its children) but the song was taken up by French soldiers under a new name: 'La Marseillaise'. On the sixth anniversary of the storming of the Bastille, 14 July 1795, it became the national anthem by decree. The painting hangs in Strasbourg's historical museum to this day, celebrating an iconic moment in its past.

On this occasion it was the river's emotional power as a redoubt – indeed, as a kind of saviour – that rose to the fore. In the centuries that lay ahead the Rhine would continue to be the troubled front line in the ceaseless agitations of France and Germany. And in de Lisle's clarion call we can hear all too clearly the tragic fervour it inspired.

'La Marseillaise' is not the only marching song inspired by the Rhine. 'Deutschland über alles' was composed for the same reasons, and at the same time. Later it was notorious as the battle hymn of Nazi Germany, but when it was written, in 1841, it was not a call for German aggression but a defensive response to *French* aggression. It aimed not to rouse feelings against foreigners but to unite Germany's own disparate parts, urging them to set aside their differences and act together in the national interest. At that time, Germany consisted of thirty-five monarchies and four republican mini-states. It was not a war cry, but a plea for unity.

A year earlier, another song emerged from the Rhineland battle front, in similar circumstances. In 1840, France's prime minister

Adolphe Thiers was leading a campaign to recover the territories on the west bank of the Rhine that had been returned to Germany after Napoleon's fall (his Rhine Confederation had been split up and handed back to Bavaria and Prussia). Now, following the failed coup by Louis Napoleon, France was again roaring with wounded national pride. Thiers demanded that the land be restored and mobilised the army. That triggered equal fury on the other side, and soon France and Germany were squaring up again.

Germany was still not unified. But unification was increasingly the word on Prussian lips. All it needed was music. And once again it was the ever-turbulent currents of the Rhine that inspired it.

A Saxony-born poet-librarian, August Heinrich Hoffman von Fallersleben, came up with the four-verse 'Deutschlandlied' in 1841, and set it to the music of a hymn by Josef Haydn. It proved popular, and in 1922 became the national anthem as 'Deutschland über alles'. In an ominous harbinger of the future, the first verse hailed the Meuse (which flows through Verdun in northern France) as being Germany's true border. The song was banned after 1945 because of its Third Reich associations, and though it was reinstated following the reunification of Germany in 1990, that first verse remains *non grata*.

Two songs, one river. France's insatiable desire to make the Rhine its border, and Germany's equally uncompromising need to possess it in full, kept making it a *casus belli*. That was the spirit in which these martial tunes – the 'Marseillaise' and the 'Deutschlandlied' – were conceived. They are fine and rousing songs, but they have weapons in their belts.

Some would say that rivers have magical powers too dark to be plumbed. But they also have creative power – the ability to nourish hearts and minds. As storytellers have known since ancient times, rivers may exist in the realm of facts, with dimensions and physical characteristics that can be measured and catalogued; but they have a vivid parallel life in the imagination, running through our thoughts as well as our hills.

In responding to the natural music of the Rhine, Wagner did much to change the way people saw it. And that new persona still makes its presence felt in dramatic ways. Every summer the Rhine Gorge erupts in an extravagant firework display known as *Rhein in Flammen* ('Rhine in Flames'). It is a tourist fiesta: the water fills with boats as flashes light the sky. With seats on the flotilla commanding incendiary prices, the music blares, the hills dance and the castles are crowned scarlet and gold.

For some visitors it might be a bit too *Twilight of the Gods*. When the *Times* columnist Bernard Levin saw it in 1987, he could only shudder: 'It was like a battle to which the opposing generals have committed their last reserves, their last ammunition, their last encouragement to their exhausted troops.'* It was, he appreciated, an effective way to echo the stormy emotions that cling to the Rhine's long history – half romantic reverie, half heroic dream. But it was also a fiery example of the way in which innocent romantic yearnings can evolve into nationalist tirades.

The Lorelei's plaintive song, after all, was also a death wish.

---

* In a journey recounted in the television series and book, *To the End of the Rhine*.

29

THE TWO TOWERS

The fact that Rome did not establish a long-lasting province east of the Rhine left its mark on European culture in many small ways. Centuries later, when Britons began to undertake Grand Tour pilgrimages to Paris, Rome, Florence and all the other citadels of classical civilisation, they tended to travel either down the Rhone to Marseille or over the Alps – no easy task: carriages sometimes had to be dismantled on one side of the mountain, carried over the pass by hand, and reassembled in Italy. But few passed through Germany. Without the centralising impetus and engineering resolve of Rome, there were no great roads through that vast hinterland. The fragmented German principalities had little interest in long-distance highways: there was no call for an Aachen to Berlin expressway, or a high-speed link between Cologne and Munich.

But Rome's absence left its mark in another enormous way. The fact that the Latin empire stopped at the Rhine, and built its camps and castles close to the waterway that kept it safe from the German hordes, meant that there were always two Germanies: a Romanised western sphere of philosophers and vineyards, and a much more war-like eastern zone. The Rhine was the ill-starred faultline, and in many ways still is.

This has been much remarked on, especially in recent years, when the differences between what used to be West and East Germany continue to propel current affairs, in sometimes vexed ways. Indeed, a learned German friend of mine has long made a point of referring to Prussia's most famous monarch as 'Frederick-the-not-so-Great' and tends to go quiet when the topic of German reunification comes up. In his eyes, unity is not Germany's natural state, or even very desirable – on the contrary, it was always an attempt to clamp together a region of two ill-fitting halves. The first unification, in 1871, was achieved by war, not peace: it was a forceful expression of gaunt Prussian imperialism, forged by Bismarck on the back of his successful campaigns against Denmark, Austria and France. Far from being part of Germany's time-honoured heritage, it was a glory-seeking dream – and one that in obvious ways led to calamity.

I did not grasp his full meaning until I went to Bingen, Hildegard's command post on the Rhine and a favoured Cold War haunt of John le Carré. There was plenty to enjoy: green hillsides cloaked in vines; the broad river swishing into its gorge; the pencil-slim barges cutting upstream with fluttering pennants; the spires of Hildegard's abbey rising above the vines and rooftops of Rüdesheim, the quaint village on the other bank. It seemed a perfect emblem of the way people used to think of Germany before the horrors of the twentieth century: as an innocent fairytale kingdom of gingerbread houses, red riding hoods and damsels waving their tresses at passing knights.

But Bingen has another unusual feature. It is a town of two towers. They stand on either side of the Rhine, glaring at each other. Both are monuments to Germany's past – but embody different ideas of that past.

The first tower, lording it over the river from its hill above the north bank, is the Niederwald, a heroic statue that honours that first unification of Germany in 1871. Completed a dozen years later, it features a 32-foot-tall female warrior symbolising 'Germania' – like France's Statue of Liberty (which was gifted to the USA), but with one hand holding a sword and the other a crown.

The inscription runs: 'In memory of the unanimous victorious uprising of the German people and of the reinstitution of the German state.' On its base we find lines from yet another marching song, 'Wacht am Rhein', or 'Watch on the Rhine'. Like 'La Marseillaise' and 'Deutschland über alles', this was written as a rallying cry – the tune that accompanied Bismarck's army as it marched into France in 1870. The plinth speaks of 'the Rhine, the Rhine, the German Rhine', and has lots of soul-stirring references to the Kaiser and the fatherland.

It was written in 1840 by a German poet, Max Scheckenburger (who grew up near Lake Constance before moving across to the Swiss side of the river). To listen to it today, it sounds remarkably like a Nazi-era marching song – and indeed it was frequently played to introduce German radio broadcasts in the Second World War. It was also the musical signal that launched the Battle of the Bulge in 1944.

In the rough-and-ready English version, it goes:

> We rise at your command, Oh Lord.
> With trust in God take up the sword.
> Heil Wilhelm! Death to every foe.
> With blood will the river flow.

Arm raised in triumph, the statue appears to shake its fist at the river over which France and Germany have clashed so often. If you

The Niederwald Monument: still keeping watch over the Rhine.

glance up first thing in the morning, with sunlight shimmering on the water, it is as if the ghosts of all those earlier conflicts are standing to attention, swords unsheathed: Roman legions, rampaging Franks and Goths, mercenaries, artillery batteries and dragoons. The tragic back-and-forth of occupation and liberation that has scarred this region . . . all seems to boil and swirl in the racing current.

Bismarck had himself commemorated in 240 columns of this sort, 173 of which still stand. All of them salute the idea of unification as if it were Germany's original or natural state. It might not be either. The Rhine's twisting history may have created a singular kingdom of its own.

~

The second tower stands for something markedly different. Not the glory of war, but the gentler arts. Hidden in the woods south of the river, near Hildegard's abbey at Rupertsberg, the Kaiser Friedrich-Turm looks more like a parish church than a war memorial. It was built following a bequest in the will of a Bingen mayor who wanted to honour Kaiser Friedrich III – the mild-natured son of Kaiser Wilhelm I who inherited the German throne in 1888. It seems likely that he would have been a peaceful and cultured monarch, since by all accounts he represented the Germany of stained glass and university libraries rather than the chest-beating *Vaterland* of stamping feet. He married one of Queen Victoria's daughters and was a friend of Prince Albert. It is widely thought he would have proved, had he survived, a calm and thoughtful figurehead. But he was felled by cancer only ninety-nine days into his reign, and the crown passed to his more abrasive son, Kaiser Wilhelm II (known in Britain as Kaiser Bill).

Bingen's twin towers thus represent contrasting aspects of Germany's national character that were laid down in Roman times. Rarely does the past stream into the present so clearly. As James Hawes has argued in *The Shortest History of Germany*, this has always been, on the profoundest level, a nation of two halves. The Romanised Rhine nourished a realm of cathedral choirs, vineyards and universities, while the trackless forest to the east, which Rome so feared, developed a more strident, martial culture based on land and family. In this analysis, the so-called 'unification' of 1871 really was a Prussian takeover.

'East Germany didn't become different because of the Russian occupation,' wrote Hawes, tracing the most recent expression of this schism, 'it had always been different.' It was not 'Germany' that produced the 'blood and iron' of Bismarck's united nation, but Prussia.

This has become a matter of keen topical interest. It has been much noticed that it was primarily Germany's Prussian voters who cheered Adolf Hitler to power in the 1930s. And today it is that same territory (the former East Germany) that is driving the *über*-nationalist AfD (*Alternative für Deutschland*) closer to significant power.

At last I understood what my friend meant when he disparaged Frederick-the-not-so-Great. As King of Prussia, he represented the army-on-the-march strand of Germany that brought drums and guns, not the cathedral spires of Charlemagne and the music of Beethoven.

The towers of Bingen are sentinels to this eternally split nature – the one a stirring monument to war, the other a quiet shrine to peace. Across the ever-pressing Rhine they stand, watching the boats cruise by.

Bingen honours its medieval beginnings by echoing Hildegard's plainchant ('Love Aboundeth in All Things') and the sound of church bells. But what does it say about modern times that the Niederwald, an overbearing monument to war, has a cable car hoisting thousands of tourists up to its glory-seeking rhetoric, plays a starring role in the tourist literature and is depicted on a thousand posters, while the more modest memorial to peace is tucked away in the woods, almost forgotten, next to a telecommunications aerial, barely visible from the town and the river?

30

THREE POEMS

For thousands of years, millions of people have worked together to create the human histories of our three rivers. Not surprisingly, since they deal in thoughts and emotions, some of the most memorable evocations of what the waters meant were composed by poets. In this brief survey we can do no more than passing justice to the whole panorama of European life that the rivers have helped to instigate and come to stand for. So let us consider three different poems, one to represent each river, in the hope that these small whirlpools might let us glimpse the bigger story of what the waters have meant to so many for so long.

RESTLESS SPIRITS

It is tempting to think that the most famous poet of the Rhone, Frédéric Mistral, must have chosen his name (which means 'masterly') as a nom de plume to demonstrate his love for the wind that whistles through his homeland – a wind that holidaymakers in Provence usually hear about in the first few minutes of any conversation. But in fact it is the other way round: the Mistrals were a well-to-do family in St-Rémy-de-Provence, south of Avignon, for at least five centuries before the poet himself was born in 1830. It was an accident of birth rather than a cultural whim that led young Frédéric to study at Aix-en-Provence and settle in Arles.

Aside from his literary gift, he was a lexicographer whose firm commitment to the cause of the Provençal language (Occitan, as in Langue d'Oc) led to him to compile its first dictionary. His most ambitious poem, *Mirèio* (or *Mireille*), was a patriotic attempt to give the language of his region a sustained piece of classical-style literature.

A twelve-song sequence, written in 1859 as a tribute to France's troubadour past (Provençal being the language of courtly love), it told the tale of two star-crossed lovers, the daughter of a wealthy farmer and the son of a basketmaker, lost in the class divide that ruled this 'empire of the sun'. He wove their story around local myths and fairy tales concerning the dragon of Tarasque (a creature with a lion's body, the head of a snake, and a taste for human flesh) and the Venus of Arles – a statue of Aphrodite unearthed in the Roman arena in 1651. The story of the Tarasque is recalled today in the town of Tarascon, a Roman-era fort on the mouth of the river where, in legend, a river monster would devour unwary travellers.

After France's pre-eminent statesman-poet Alphonse de Lamartine hailed *Mirèio* as a national epic in the style of Homer and Virgil, Mistral won the Nobel Prize for Literature in 1904. He used the prize money to create a museum of Provençal life in Arles, and there's a statue of him nearby, sporting a hat, leaning on a stick and holding a coat over his arm with the innocuous air of a commuter waiting for the Lyon train.

The work that concerns us here, however, is his 'Poem of the Rhone' – his 1897 tribute to the river that ran through his youth. Ironically, since his purpose was to preserve and promote the language, the poem found a wider audience by being translated into fifteen foreign tongues. But even in translation its heritage was unmistakable. Ever since the eleventh century Occitan had been home to the troubadours whose ballads provided the Middle Ages with much of their musical heritage.

The traditional love song also owed something to Provence's natural openness to ideas from elsewhere. The Rhone opened a door not just to Rome, but also to the music of Andalusia and Moorish Spain. The 'Poem of the Rhone' (*Lou Pouemo dou Rose*) was, like *Mirèio*, a salute to that.* It narrates the romantic adventure of two young lovers as they board a barge in Lyon, part of a flotilla (a hundred vessels, a thousand horses) embarking on a boisterous trip south to Beaucaire, near Arles.

This was the crossroads of Roman Provence – the place where the river broke through the hills and where Gaul's first Roman road (the Via Domitia from Italy to Spain) crossed the Rhone. In 1453 Beaucaire was given royal assent to host an annual fair – one of those Provençal fiestas involving bullfights, plays, fortune-telling, fisticuffs, wine, food, music, dance and every other sort of naughty pleasure. (Times have changed. In 2016 the mayor of Beaucaire, impressed by Britain's withdrawal from the EU, named one of the town's roads the Rue de Brexit. According to the *Washington Post*, it was an 'ugly dead end' in an industrial estate.)

The leading character was a ferryman's daughter, Anglore, who (like the Rhine's Lorelei) represented the magic of the Rhone – the river that 'with thy name alone charms the world'. On one level Anglore *is* the 'flower of the Rhone', but her true feelings, as she sifts the waters for gold, are reserved for a spirit named the Drac (a

---

* Its success can be seen in the way that the French *chanson* continued to thrum in later years, in the music of Charles Trenet, Edith Piaf, Jacques Brel and many others.

cousin of the Tarasque and perhaps a distant relative of Dracula) who lives in a grotto on the riverbed, inveigling passing maidens into his icy embrace. He is a shapeshifter, and Anglore is lulled into believing that the youth on her barge is this demon in disguise; in fact he is a Prince of Holland.

The lovers' ill-fated passion leads them through vineyards, olive groves, castles and chapels, trailing their fingers in the flood. But the Drac, jealous of Anglore's love for this Dutch arriviste, orchestrates a tragic climax. The Rhone's first steamboat (symbol of the modern age) crashes into them, spinning them to their death in its ever-rushing waters.

It is hard not to see the forward march of these events, and the clouds that gather over the lovers, as echoing this unstoppable current. But the Mistral – the tree-bending wind that rips down the Rhone for days at a time, rattling windows and driving people out of their wits – has a part to play, too. A natural effect, it occurs when high pressure in Biscay or the North Sea coincides with low pressure in the Mediterranean, squeezing the Atlantic air and blowing it south. The fact that it comes out of a cold blue sky makes it feel irrational and unsettling: a dementor that has bent this region to its whim. Petrarch's famous Mont Ventoux owes its name to the gale that whips along its peak, and the typical *mas* or farmhouse of Provence tends to face south in order to present its back to the gale. Church steeples have vents to let the wind through, while villages huddle round squares against the common foe.

This restless spirit is ever-present in Mistral's poem. The characters are swept along by forces beyond their control – the river and the wind. Like figures in an Attic tragedy, they sing love songs even as they drift towards the whirlpools. There is nothing of industry and commerce in this vision: this Rhone is raw nature.

ROMANTIC LONGINGS

When Byron first saw 'the wide and winding Rhine' in *Childe Harold's Pilgrimage* – 'a work divine / A blending of all beauties, streams and dells / Fruit, foliage, crag, wood, cornfield, mountain, vine' – he was, as we have seen, prompting thousands of Britons to follow in his footsteps. Many of these tourists were also poets, and they tended to see the Rhine through the same Romantic eyes to an extent that became formulaic ("Twas morn, and beauteous on the mountain's brow . . .', etc.).*

But the most persuasive poet of the Rhine is Friedrich Hölderlin, the idealist who, though no Rhinelander, spent thirty-six years gazing at the Neckar (which meets the Rhine near Heidelberg), meditating on the symbolic power of water from the tower at Tübingen where he was a schizophrenic recluse. His 1808 poem 'The Rhine' is a biographical sketch of the river – from its beginnings as an energetic *Jüngling* sprinting through the Alpine 'castle of the Gods', to the 'kinglike spirit' it goes on to become. Actually, in the poet's mind it was not the river's destiny that counted so much as its desperate and continual urge to escape. Even in infancy it was 'a youth clamouring to be set free'.

Throughout, the Rhine appears as a figure out of Greek mythology.

And the wrath of the young god was fearful
As he tore at his chains in the dark.

---

* Much the same Romantic frisson informed the prose of the Swiss-German novelist Hermann Hesse, who asked, with reference to his broken-personality Steppenwolf: 'Who read by night above the Rhine the cloud script of the drifting mists . . . ?'

Thereafter the river slows, maturing into the sort of companionable presence that could inspire the idealistic dreams of a Rousseau, a Byron . . . or a Hölderlin. It becomes 'Papa Rhine', a tolerant presence. But in so doing it also becomes the symbolic link between the Olympian realm and the everyday human world. Or, as the final lines of the poem put it:

> When the day draws to a close,
> The divine artist, best of our gods,
> Inclines to our earth.

In giving the Rhine this supernatural role Hölderlin did not fail to notice that it was one of three streams to spring from the same rocks, and was thus part of a wider European pattern. He added that this 'noblest of all rivers' became itself only when it 'broke free from its brothers / the Ticino and the Rhone'. And though at first it ran east towards Asia, it was soon answering the siren call of Germany and turning north. In genteel old age it remained a radical force. Its murmur, as it flowed 'quietly and happily / through German fields', was 'the voice of freedom'.

That allowed Hölderlin, an avowed Romantic, to see the river as linking the divine to the human. In today's terms we might translate this idea to suggest that it also links the past and present. And if we follow the Rhine downhill from Switzerland to the Dutch border (past that historic run of great cities: Basel, Strasbourg, Freiburg, Speyer, Mannheim, Worms, Mainz, Bonn, Cologne and the rest) we feel another connection. It is a geography field trip, but also a course in German history.

WELCOME TO PURGATORY

It is no surprise to find that in his travels Byron wrote some 'Stanzas on the Po', which used the water as a *mis-en-scène* for his own hopes:

> River, that rollest by the ancient walls
> Where dwells the lady of my love, when she
> Walks by thy brink, and there perchance recalls
> A faint and fleeting memory of me.

The river as fickle mistress . . . it might not be Byron's greatest work. But Edgar Allan Poe enjoyed the pun on his own name sufficiently to write a similar verse under the deliberately coy title: 'To the River'.

> Fair river! In thy bright clear flow
> Of crystal wandering water,
> Thou art an emblem of the glow
> Of beauty . . .

Romanticism, we might think, reading these ho-hum rhymes, has a lot to answer for. But Byron and Poe were just passing through. For a more bracing consideration of the Po and its place in the world we need to consult the greatest poet of Renaissance Italy, Dante Alighieri.

Born in Florence in 1265, he was a stout citizen who bore arms in his home city's defence at the Battle of Campaldino in 1284, but was banished in the civil war that flared between Florence's rival factions, and thereafter lived as an exile in Rome, Siena, Verona and Ravenna. It is fitting, then, that his greatest work, the *Divina*

Commedia or *Divine Comedy*, should be an epic journey. Led by the poet Virgil, Dante makes his way through the underworld, beginning in the fires of Hell before visiting Purgatory and the long-promised pastures of Paradise.*

One of the first people he meets in the second circle of the Inferno is a woman named Francesca who has been damned (like Lancelot, whom Dante mentions) for being caught in the 'carnal sin' of adulterous love.

She begins by telling him where she is from.

> The place where I was born is by the sea
> Where the River Po runs down to rest in peace
> With his attendant streams . . .

This is usually taken to mean Ravenna, where Dante himself settled. A few pages later he meets a woman who is enduring a cruel punishment for presuming to foretell the future: she must walk with her head turned backwards for the rest of time. Her name is Manto, and it is in the marshland of the lower Po that she finds a swamp in which she can suffer alone. But enough people join her until they have founded the city of Mantua, protected by pestilential waters – a place of no little significance in the context of our river journey, since Mantua happens to have been the birthplace of Virgil, Dante's guide and mentor through the ways of Hell.

Then, in Book 14 of the *Purgatorio*, Dante hears the lament of a man denouncing the degenerate ways of those who live on the Arno (a pointed accusation aimed at Florence, Dante's home city), and

---

* It is worth remembering that the word 'divine' was not part of the original title. It was an adjective added by so many of the work's admirers that it became first attached and then embedded, as if it had been Dante's idea all along.

contrasts that area with Romagna, the virtuous realm between the mountains and the Po.

Finally, in the *Paradiso* (the story is framed as a redemptive ascent to salvation) we find a remarkable passage in which the poet meets the shade of the great Emperor Justinian, who recalls the mighty ages of the past by recounting the flow of events that have occurred on their rivers. The Po, the Rhine and the Rhone are all mentioned, as is the Rubicon.

These are not casual allusions. Earlier, when Virgil led Dante to the centre of the Inferno, they came upon a terrible sight: a river of blood:

> On we passed
> To where there gushes from the forest's bound
> A little brook, whose crimsoned wave yet lifts
> My hair with horror.*

Nothing they had seen so far, Virgil says, was 'so worthy of regard'. The blood-stained water, he explains, was made from the tears of a giant locked away in the rocks of Mount Ida, in Crete – the earthly surrogate of Mount Olympus, home of the quarrelsome Greek gods. All the rivers of the underworld were made from these tears, seeping through fissures in that tormented mountain. All were fed by the same fountain of woe.

This is only one of the ways in which the symbolic power of flowing water runs through the *Divine Comedy*. The path to Purgatory is blocked by the Lethe, in which penitents must wash away all

---

* This is the slightly archaic 1804 translation by Henry Francis Carey. There have been dozens of other versions in the years since.

memory of their sins (in the cave of Hypnos – as in 'hypnotism') and then by the Eunoe, which does the opposite: intensifying the memory of past virtues.

Thereafter the pilgrim encounters the other rivers of the underworld: the Styx, in whose enchanted water Achilles was dipped by his mother, Thetis (who short-sightedly held on to his heel); the Acheron, another flood over which souls must also be rowed to enter the underworld; the Phlegethon, the river of fire; and the Cocytus, which marks the outer limit of Hades and is also a river of lamentation – an ocean of sighs.

But that trickle of bloody water out of Mount Ida in Crete is the origin of all these, and many other rivers too – which means that all the water that flows through the known and unknown world is made of tears. In a trim foreshadowing of modern anxieties about climate change, Dante meets the spirits of those who have 'done violence to nature' before proceeding to the centre of Hell, where historic traitors such as Cain and Judas suffer eternal punishment neck deep in frozen water, or held face down on the ice, eyes frozen to the unforgiving ground.

At the centre of the flood, waist deep in the frozen water, stands Satan. This is the place to which he has fallen, and it is very like a glacier.

But Dante saw the river not only as an emblem of death and despair, but also as the path to hope and love. The primary feature of the 'Earthly Paradise' at the end of the poem is a stream of pure, bright water.

> Incredible how fair: and from the tide
> There, ever and anon, outstarting flew
> Sparkles instinct with life . . .

~

On the surface, the rivers reflected in these three poems have strikingly different characters. Mistral's Rhone is an anthem of doomed love; Hölderlin's Rhine an operatic blast of freedom; Dante's Po a thread through a more intimidating cosmology. But beneath these distinctions they have something in common: a split personality. Just as rivers divide but also connect, so these poems flicker constantly between images of good and evil, dread and hope. Their banks contain the best (as Matthew Arnold put it) that has been thought and said – but also the worst.

Perhaps that is why Virgil said, as he led the way to the tearful origins of Europe's rivers, that he was giving Dante a tour of 'human history'.

31

THREE PAINTINGS

The wealth siphoned out of our three waterways permitted a scintillating range of patrons – emperors, kings, bishops, tycoons – to assemble treasures that are now the basis of superb public collections of art. Cologne, Basel, Lyon, Arles, Milan, Venice: all have world-class galleries – everything from medieval and Renaissance masterpieces to Cézanne, Monet, Picasso and the rest.

The rivers themselves have featured in thousands – or millions, if we include enthusiastic amateurs – of such works. But to explore their role as muses let us choose just one for each stream. Perhaps that will help us see their different natures: one as an instrument of war, another as a stage for dreams, and the last as a place where nature rubs shoulders with nurture.

THE BARRIER

In 1860 Wilhelm Camphausen, Professor of Painting at the Düsseldorf Academy and a noted composer of military scenes, unveiled a new work, *Blücher Crossing the Rhine*. Painted at a time when Bismarck and his followers were agitating to forge a German union (under Prussian leadership), he chose to capture the historic moment when Prussia's Field Marshal Gebhard von Blücher followed up his victory over Napoleon at Leipzig in 1813 by pursuing him across the Rhine at Kaub.

The French army, still bleeding from its disastrous campaign in Russia, had lost half a million troops, and the emperor was facing ruin and exile. As we have seen, Blücher's 3rd Silesian Army reached the river in December, and threw a pontoon together in a matter of days.

It was New Year's Eve, and an important symbolic moment. It marked the end of Napoleon's empire on the Rhine and announced Prussia's arrival as a 'Great Power'. Blücher was depicted in the grand style, on a gleaming chestnut horse, its nostrils billowing mist as a bright sun lit up the early-morning snow.

The marshal has his finger raised skywards in the middle of a festive crowd. Troops warm their hands on a twig fire, children throw snowballs, a dog dances a jig. An ice-blue sky gleams over the Prussian Lancers leading their mounts to the water's edge. In the background the army is already on the bridge to Pfalzgrafenstein, the fortified island in the middle of the stream. Twin castles stand on the crags like eagles.

It is the romantic Rhine in full battle dress, a wartime tableau filtered through a carnival atmosphere that resembles the famous Frith painting of Derby Day in England, produced just two years earlier. And though war was not quite an annual fixture in this part of the Rhine, neither was it rare. As we know, the list of such crossings, starting with the tidal surge of the barbarians over the Rhine, is as long as the river itself.

THE DREAM

In February 1888 Vincent van Gogh left Paris and went south to the old Roman city of Arles, on the Rhone just north of Marseille. The warmth and light didn't do anything for his mental health – nine months later he would chop off his ear with a razor and be confined in an asylum – but from an artistic point of view it was exhilarating. He made 200 paintings, some of them among the most valuable in the history of art.

Starry Night over the Rhone is one of several nocturnal visions rendered in vivid swirls of blue and yellow. In Provence, these colours became his palette. Yellow was the colour of sun on parched earth, the rich ochre gleam of wheatfields and sunflowers, and the shade of his house on the Place Lamartine. Blue was the Mediterranean sky and the Rhone – even at night.

In this midnight-blue vision, cartoon stars blaze out of the heavens and blend with the gas lamps of Arles to fill the water with golden drops. It is not an accurate depiction of the sky: the big dipper has been shunted out of its orbit to occupy centre stage – in real life it would be behind the painter's back, not glowing in the south. But details like that serve only to confirm that this is not a literal piece of work. Van Gogh attached candles to the brim of his hat, worked through the night like a man with his heart (or hair) on fire, and painted not what he saw but what he felt.*

'I lose myself and paint as in a dream,' he wrote.

The dream in which he lost himself was one of heightened colour. In a letter to his sister he described the stars as 'lemony'; to

---

* Modern art tourists can follow the Van Gogh tourist trail to the exact spot overlooking the Rhone where he set up his star-gazing easel.

his brother he added: 'The night is much livelier and the colours more intense.'

While Van Gogh escaped into himself by pouring out his feelings about the landscapes of Provence, the world was spinning. Slavery was finally coming to an end in Brazil (forty years after its abolition in France, and seventy years after Britain). Germany was having its 'year of three emperors' (after Wilhelm I died in March 1888, the brevity of Friedrich III's reign, commemorated in the tower at Bingen, meant that Wilhelm II was the third monarch to be crowned that year). London was being haunted by Jack the Ripper. France was busy annexing Polynesian islands – a topic that might well have engaged the attention of Van Gogh's friend Paul Gauguin, who was staying with him in Arles but would shortly be heading for Tahiti. Paris, meanwhile, was a cacophony of hammer blows as the Eiffel Tower rose above the rooftops. The Riviera was still recovering from an earthquake the previous year.

The previous year – 1887 – was the year of the so-called Schnaebelé Incident – a crisis that flared dangerously when a French policeman in Alsace was arrested by his German counterparts, sparking a week of sabre-rattling that ended only when Bismarck (who may or may not have ordered the arrest) personally interceded to have the man released.

In a world as jittery as that, the starlight sparkling in the Rhone in the deep blue of a Provençal summer must have seemed astounding – a river of visions.

THE WATER FEATURE

While the Rhine and the Rhone are dashing characters – forthright enough to make lively artists' models – the Po is an altogether muddier affair. It gleams through some Turner watercolours, swirling beneath misty apricot skies, but these could just as easily be portraits of the Rhine, or indeed Ullswater – his own vision predominates. Apart from that, it has not featured in as many memorable works as its German and French cousins.

This may have something to do with the fact that it runs through flat, featureless countryside, without the cliffs and gorges that might lend grandeur to such scenes. Its cities do not bestride the current, and it is famous for its fogs. It carries a milder charge than the other waters. Even the name is a giveaway. The Rhone and the Rhine are derived from their Gaulish names: Rodonos and Renos, both meaning 'fast-flowing'.* Their dynamism is patent and obvious. The Po, in contrast, blends the Latin words for 'plain' (*padano*) and 'swamp' (*pollicinum*) – the marshy delta is known as Polesine to this day. Far from being the swashbuckling hero of the terrain through which it passes, it slinks along in the background. It is clearly visible as a slender blue line marching behind the armies in the tapestries that commemorate the Battle of Pavia, the final major encounter of the Italian War between France and the Habsburg Empire in 1525, but even there it is somewhat incidental to the main action of the scene.

In *Landscape with Merchants*, painted in the seventeenth century by the French artist Claude Lorrain, the Po is in the foreground, but barely recognisable. Idealised and generic, this is not a portrait of a place so much as the presentation of an idea. The river, the boats, the trees, the mill, the ruined tower, the human figures, the flowerpots, the bright sunset, the suggestions of Roman mythology: these are standard-issue ingredients.

---

* In Italian the Rhone continues to be known as *Il Rodano*.

Lorrain himself embodies the tangled roots of this image. Born Claude Gellée in the Duchy of Lorraine in the early years of the seventeenth century, he was a French-speaking subject of a German empire who fled to Italy when the Rhine was threatened by Cardinal Richelieu's territorial ambitions during the Thirty Years War. He spent the rest of his life in Rome, far from these horrors, becoming in effect a one-man confluence of our three streams: French, German and Italian culture flowed through both his mind and his brush. What he painted above all was tranquillity – a radiant refuge from a world in turmoil.*

That his work then took a twisting journey of its own is clearly visible in a British engraving of his original painting: *A View on the River Po in Italy*. It was published in Georgian London in 1769, when a printer named John Boydell invited the engraver James Mason to make a copy of it for his new venture, *A Collection of Prints Engraved After the Most Capital Paintings in England*. Hogarth, Poussin, Rembrandt and Tintoretto were all in there, but Claude was pre-eminent: eight of the book's 110 images were renderings of his bright and polished landscapes.

Claude was far and away the most admired artist in Britain at that time – no stately home was complete without one of his paintings or, failing that, a print.† When London's National Gallery opened in 1824 (in a townhouse in Pall Mall), he was the best-represented artist in the entire collection, and as such was a major influence

---

* His compatriot Jacques Callot went to Rome too, but returned to Nancy to make the horrifying engravings of the war-torn Rhineland – farms ruined, churches burned and prisoners butchered – all published as bitter commentary in *The Miseries of War*.

† Chatsworth, seat of the Duke of Devonshire, had 195 of them – done by Claude in a survey of his collected works known as the *Liber Veritatis*. It is now in the British Museum.

on Gainsborough and Turner. Constable called him 'the most perfect landscape painter the world ever saw'. But his true influence ran deeper: his distinctive rendering of rural beauty stirred with antiquarian echoes – it was in the Po that Phaethon, the son of the sun god Helios in Greek mythology, was supposed to have drowned when he lost control of the chariot containing the sun and fell into the river. His work did a huge amount to mould England's sense of what landscape meant.

Here was born the native style of English pastoral whose Latin origins made it 'Augustan'. It produced the raked 'heroic' couplets of Gray, Pope and Dryden, but found even more refined expression in the country's aristocratic parkland. The artfully conceived estates of landscape architects Capability Brown and William Kent – complete with lake, woodland glade, ha-ha, sheep pasture and Roman grotto – owed everything to the classical mood of Claude's sylvan arcadia (nicely exemplified by Mason's engraving of his Po view). 'Nature' came to mean Palladian lines in an English meadow, accessorised with a temple to Diana, a Doric column, a fountain to Cupid or a cow nuzzling at the edge of a pond.

After all, the whole idea behind 'picturesque', a concept born at just this time, was that the natural world should resemble a picture.

That explains why Boydell and Mason's version of Claude's Po feels so like Stourhead or Blenheim Palace. It has been domesticated. The rakish prows of the boats do have a certain Venetian-gondola swagger, but in most other respects this resembles a secluded country estate. In this first spring of the Grand Tour, the beauties of classical Italy were sacred treasures, but nature had not yet been elevated into 'the sublime'. This was a picnic spot: something to be seen, sketched and hung on your orangery wall. Far from capturing anything specific or essential about the river, Mason was content to depict it as a rest stop for a travelling marquis, a sojourn in idyllic countryside. The Po was literally being pressed into service as an advert for the idea of getting away from it all.

The chosen format – the engraving – was the height of fashion, too: printing had advanced wonderfully since the days of Gutenberg and Caxton, and Boydell was enjoying such success with his prints of famous art (notably with his Shakespeare series) that he was single-handedly reversing Britain's balance of trade in such works with Paris – an achievement that led to his becoming mayor of London.

John Ruskin would later denounce Claude Lorrain as dated and escapist, and he had a point. It is all too clear to modern eyes that his neo-classical tributes to rural repose, however perfect, were to some extent inspired by a desire to draw a veil over the origins of the wealth that made such visions possible. And though there was nothing scandalous about his mythological scenes – Aeneas landing in Rome, Venus having a bath – the images were suffused with a polite sense of ownership: this was land that could be possessed as well as tamed.

Mason's rendering of Claude's Po is in this way exemplary. It is not a river of power or visions, but a river re-imagined to symbolise rustic charm.

~

The nation-shifting power of the Rhine . . . the hot dazzle of the Rhone . . . the timeless serenity of the Po . . . In their Alpine infancy these rivers share the same splashing and exuberant energy. But below their lakes they are different characters. For Van Gogh the Rhone is a rush of intense feeling. In Camphausen's eyes the Rhine is both a frontier and a military asset, a barrier and a spur, the cause of conflict and the bridge to peace. And in Claude's luminous and wistful depiction, the Po is what *all* rivers aspire to be – benign patches of nature ripe for grooming by human hands.

There is truth in all these renderings. The Rhine did indeed become Europe's power cable. The Rhone embodied its dream of infinite summer. And there have always been reflections of Greece, Rome, cardinals and artists in the slow, mirrored waters of the Po.

32

RIVERS OF BLOOD

As daylight faded over the East Anglian coast on the evening of 30 May 1942, the air quivered with the thunder of planes. At first no one paid it much attention: there had been lots of RAF bomber sorties over these parts in recent weeks, so the noise was not unusual. And it was pelting with rain, so there was nothing to see.

But there *was* something different in the air that evening. The rumble in the clouds went on . . . and on . . . and on. In the end it took two hours for the noise of the engines to drone away to the east, long enough for even the most distracted observer to wonder what it meant.

It was the opening salvo in a dreadful campaign. War had returned to the Rhine. And though the technology had changed, and there was a new cast of characters, it was the same old story. The cities on the river that had for so long been the demarcation line between Franco-Roman and German Europe were about to be bombarded all over again.

Arthur Harris, the Great War veteran in charge of Bomber Command, was itching to land a decisive blow on Nazi Germany – in part to please Stalin, who was begging for action in the west, and partly to avenge the 1940 Blitz on London. His idea – not a subtle one – was to assemble the greatest armada of airborne firepower the world had ever seen, and send it off to incinerate one of Germany's most famous cities: Cologne.

It was 'area bombing', and Winston Churchill supported it, if only to demonstrate that Germany was not invulnerable and he was not beaten.* For his part, Harris did not even attempt to hide his feeling that it was an act of biblical vengeance. 'They sowed the wind,' he said, quoting the biblical Book of Hosea. 'Now they are going to reap the whirlwind.'

The assault involved co-opting (and risking) almost every bomber in the country, using fifty-three different airfields. Take-off times were synchronised to create a 70-mile convoy over the North Sea to overwhelm Cologne's defences through an elaborate flight plan that involved attacking the target from four different angles.

As chance would have it, the planes unloading their bomb bays over a symbolic German capital were named after *English* cities – Halifax, Manchester, Lancaster.

The operation was codenamed 'Millennium', and when the one-thousandth plane was signed up, on 30 May, the green light flashed.

---

* Actually the idea of bombing German cities came from a 1942 paper written by Professor Frederick Lindemann, Winston Churchill's German-born friend and adviser. An ardent anti-Nazi, Lindemann delivered his paper to Cabinet, emphasising that maximum casualties should be sought. The Blitz had not dented Britain's appetite for war, and the idea that aerial bombing would shatter German morale won the day.

By the time night fell, a few dozen more had been found, so in the end there were 1,046. But the apocalyptic name remained untouched.

It is common now to see the blanket bombing of Germany's cities as a crime. Even as we remember the scale of the provocation, and salute the courage of the pilots who lost their lives (many of them pale recruits barely out of their teens*) the tactic sits badly with modern sensitivities. Industrial-scale assaults on civilian populations seem barbarous, however grave the context and however many stories they generated (the great Australian cricketer Keith Miller, for example, who took part in those raids, made a point, when flying over Bonn, of whistling Beethoven, whose birthplace was being mauled down there).

But the raid on Cologne differed from the horrors that would come later, in 1944, when cities such as Hamburg (Operation Gomorrah – 50,000 killed) and Dresden (death toll: 100,000) were incinerated. The purpose then was to horsewhip a beaten opponent and perhaps hasten the end; but in 1942 the picture was less clear. The argument for a distraction to relieve the pressure on Stalin had merit – this was the year in which Germany was tearing through Ukraine en route to Stalingrad. And there was a military motive as well. As Churchill later wrote, it was hoped that the bombing of Cologne would alarm Germany into ordering its factories to make fighters, not bombers, and thus reduce the risk of a second Blitz.[†]

---

* Guy Gibson, the squadron leader in charge of the Dambusters raid, was just twenty-three.

† 'This was the beginning of defeat for the Luftwaffe, and a turning point in our struggle for air supremacy . . . By forcing the enemy to concentrate his strength on defending inner Germany the Western Allies gained the complete air superiority they needed for the approaching cross-Channel invasion.' Winston Churchill, *The Second World War*, vol. 5.

Operation Millennium still feels shocking, however – a grim attempt to rain terror on an old and cultured city. The planes dropped explosives to blow off the rooftops, followed by incendiary bombs to burn the rest.

Thus was a historic city, home of the relics of the Magi, cremated.

The Rhine had a specific role to play – that of a navigational aid. There was no satellite positioning in the Second World War. The RAF bomb-aimers had to rely on maps, compasses, the stars and dead reckoning. But thanks to the river, Cologne was not hard to locate. That great ribbon of water was visible even on a dark night, and on 30 May the sky was clear, so the Rhine reflected the moon like a scimitar leading straight to the heart of the city. Thanks to the bulk of its cathedral (avoided by the bomb-aimers to preserve it as a directional aid) it was nigh on impossible to miss.

In theory Cologne could dip its fire hoses into the river and pump the ever-bountiful Rhine onto the flames. But nothing could suppress the inferno that night. If the Allied pilots had turned around as they rattled over the Dutch coast (150 miles away) they would have seen Cologne writhing in a hellish red glow. In fact the death toll was light– 'only' 500 perished. But thousands more were injured, and 20,000 houses were burned to the ground. In a single night, a great city was reduced to ash.

In the years to come all the river's cities would suffer this fate. There were four anti-aircraft guns on the Nibelungen Bridge in Worms, not nearly enough to deter the Allied raids. One, in March 1944, involved 300 planes and destroyed a third of the city's buildings. The bombs missed the bridge, but the Germans blew that up as they retreated.

Cologne's cathedral: miraculous survivor of a thousand bombardments.

The Rhine had never stopped being a battlefield. As we have seen, the fact that France wanted it as its frontier, while Germany had an existential need to possess the western shore, had been the cause of endless Franco-German friction. The two most recent conflicts – the Franco-Prussian War of 1870 and the First World War – had left France more determined than ever to make the Rhine its border.

The first of those wars had been humiliating, in that France was toppled in weeks by a fledgling nation only a fraction of its own size, and the horrors of Verdun were still raw. Hence the Maginot Line, the thick row of bulwarks, tunnels and gun emplacements that ran up the Rhine from Switzerland to Belgium.

But in 1936 Hitler walked unopposed into Alsace-Lorraine, and the old emotions flared again. And as the war that followed approached its climax, the Rhine found itself again in the line of fire.

Not for the first time much of the fighting was over the need to seize the crossings. The most momentous of these was the failed assault at Arnhem in 1944 (the famous 'bridge too far'). It was the biggest aerial assault of the war – more than 1,500 planes dropped 10,000 men into the Netherlands that night, while armies crossed the river by raft. Work on the bridge began at 9.45 a.m., and by teatime that day thousands of trucks were rolling over the river on a broad road that rested on ninety-three pneumatic floats.

Where Arnhem failed, Operation Plunder succeeded. This was the final advance that took the Allied armies into Germany. In most minds the war was as good as won, and London's National Army Museum still calls it 'The Forgotten Battle'. It could not have felt like a fait accompli to the soldiers involved, but to the wider public the daily news was no longer a source of dread and terror.

The Germans were demoralised and gave up fast. And when the Allies crossed the Rhine at Wesel, they saw why. The bombs had done their worst: another elegant city was now a pile of rubble.

One of the first men across was Winston Churchill. After enjoying lunch with General Eisenhower in a nearby chateau he had himself taken to a house overlooking the Rhine, and insisted on being given a boat (there were news cameras present). When his party neared the bank, shots were fired and they withdrew. But it made for some excellent photos.

This crossing of the Rhine was the last major battle of the war, in Europe at least. On 3 March a tank brigade crossed the pontoon bridge, and less than a month later it reached Berlin. When the river fell, so did the country. As General Patton had said when he captured the bridge over the Rhine at Remagen (by luck as much as nerve – the demolition charges failed to explode): 'During world history, wars have been lost because rivers had not been crossed.' His words are exhibited today in the baroque-era inn Zur Stadt Mannheim, where Blücher himself stayed.

The remains of the bridge at Remagen are still in place, blackened stumps on the riverbank between Koblenz and Bonn. It was originally built as the Ludendorff Bridge to carry troops to the Western Front; it was one of the crossings over which Hitler's soldiers streamed in 1936 to reclaim their lost lands. But ten days after Patton went over it in 1945 it collapsed. Its abandoned turrets now house a Peace Museum, which has the banner left by Patton's engineers to greet any detachments that might follow: 'Cross the Rhine with dry feet, courtesy of 9th Arm'd Div.'

It is a gaunt reminder of the extent to which this ever-churning river has above all else been a wonderful machine for the manufacturing of history.

~

The Rhone and the Po were not battlefields to the same intense extent. But they did not escape unscathed. The Po, in particular, was the scene of a major offensive at the end of the war, when Allied forces were pushing north from the beachheads at Salerno and Anzio to attack Germany's 'soft underbelly'. The decisive thrust was achieved in the summer of 1944. A US Intelligence officer named William 'Memphis Bill' Mallory noticed, as he stared at his map, how relatively few bridges there were over the Po. He counted only twenty-four – eight of them for rail.

The way forward seemed obvious: destroy them.

The Allies were not concerned about securing crossing points of their own. They would cross that bridge (as the saying goes) when they came to it. The more pressing problem was Germany's ability to supply its army in Italy. Without those bridges it would be impossible.

Operation Mallory Major, a three-day aerial attack, sent sixty sorties from Corsica along a 200-mile stretch of the Po. As at Cologne, the river was a perfect aid for the bomb-aimers. At the end of the seventy-two-hour assault, twenty-two of the twenty-four target bridges had been hit, a better result than even Mallory had hoped for. The German presence in Italy collapsed. It became a double victory when the Germans fell back to the Po. They had hoped to use it as a defensive wall – the 'Gothic Line'. But the broken bridges trapped 100,000 of them on the south shore, where they were taken prisoner by the Allied advance.

Mallory was rewarded with a medal and a trip home. But in an agonising dénouement, the plane taking him home to a hero's welcome crashed on take-off, and he died. Yale University, where he had been a football star, put his name on an athletics scholarship and a gymnasium. The rest of us forgot him.

～

The Rhone saw action only a month later. Operation Dragoon (a diversion planned as a support act for Operation Overlord in Normandy) involved the landing of half a million men on the Côte d'Azur in August 1944, in the hope that this would stretch German resources in France.

It began in slightly cartoonish fashion when a detachment of marines slipped ashore on the near-tropical island of Port Cros, off Hyères, on a mission to disable the gun batteries that commanded the sea. It was well planned, bravely done – and entirely pointless. The saboteurs did all that they were supposed to, forced the Germans into the castle and set about spiking the guns – only to find that they were dummies.

The whole coast, it turned out, was barely defended. The invasion force – half a million men, with 1,400 artillery pieces and air support (from Corsica) – poured onto the beaches of the Côte d'Azur.

To this day, holidaymakers in Saint-Tropez are surprised to find themselves walking past a monument to the landings, not far from one of Europe's most golden sun traps. The inscription reads:

15 AOUT 1944
DEBARQUEMENT DES ARMEES ALLIEES.
SOUS LE COMMANDEMENT DU GENERAL PATCH*

---

* There is another such plaque at Rayol-Canadel, closer to Toulon.

The Germans mustered in Marseille before fleeing up the Rhone, aiming to make a stand near Dijon. But the Allies sent a flanking force on a parallel route, which managed to reach the sugar-and-nut town of Montélimar before the retreating Germans. There was a four-day exchange of fire, but both sides were anxious to avoid a fight, and the Germans slipped past. They raced north to Lyon, aiming to branch east to the Rhine, but General Patton's advance units were waiting for them.

In just a few weeks the whole of southern France – the Rhoneland – was freed.

All three rivers have been notably peaceful since then. For how long, no one can say. If the past is anything to go by, perhaps war will come soon. In the eighty years since the Second World War, this part of Europe has enjoyed a far longer stretch of peace than usual – the longest period without a conflagration since Roman times. But now there are other threats to contend with, such as industrial pollution, climate change and environmental decay. It is only natural that the rivers should be on the front line of these challenges, too.

33

MELTDOWN

Shortly after lunch, on 9 February 1999, the municipal authority of Chamonix, at the foot of Mont Blanc in Haute-Savoie, held an emergency conference. There had been a rapid build-up of snow in the rocky bowls above the town, and the authorities needed to discuss what to do. The evacuation of residents was mentioned.

It was already too late. Before the councillors had even sat down, a 30-acre slab of snow, a couple of feet deep, was losing its grip on the ice. Soon it was slithering down at 60 miles per hour, fast enough to demolish everything, a frozen battering ram of trees and stones. When it smashed into the village of Le Tour, a few miles up the valley from Chamonix itself, there was nothing anyone could do: twenty-three homes and dozens of barns were flattened in the time it took to cry for help.

The thump was hellish, but the silence that followed was worse. In the space of a moment, the village was pulverised; twelve people lay dead.

At a court hearing in Chambéry, in 2003, the mayor of Chamonix, Michel Charlet, was sentenced to three months in prison for what was deemed criminal negligence in not evacuating the valley sooner. By then the public mood wanted to emphasise that mayors had weightier duties than making the valley safe for snowboarders. But in truth it was beside the point to blame any one person. There was a deeper cause.

Just as the debacle in the Val de Bagnes in 1816 had created new sciences such as glaciology and climatology, so the calamity in Chamonix triggered a major rethink. The ground above the town was no longer stable, nor were the old certainties.

Two weeks later, a 150-foot wall of snow, weighing 170,000 tonnes, ripped out of the mountains and engulfed the Austrian resort of Galtur, in the Tyrol, killing thirty-one and injuring many more.

Now the sirens were sounding even louder. Two calamities in the same month? It seemed that the earth could not be trusted.

~

More than two decades on, the jury is no longer out. As John Neal, the Chief Executive of Lloyds of London, told the *Financial Times* in 2024, 'You'll never find an insurer saying, "I don't believe in climate change."' Rising temperatures are not just provoking torrid weather events; they are altering the composition of the ground itself. In the summer of 2022 the snow line – the altitude above which everything freezes – rose to 14,000 feet, only a thousand feet below the summit of Mont Blanc. The standard level in recent times has been around 10,000 feet.

The consequences are plentiful and well documented. The ice is dying. The glaciers are in full retreat, and rockfalls are common.

Geomorphologists with laptops were starting to place monitors all over the Alps. Their findings didn't amount to the kind of stories that make headlines (until it is too late) but they were troubling. In July 2023 a Swiss weather balloon found that the temperature contour known as the 'zero-degree line' had reached 17,000 feet – an all-time record that was some way higher than the summit of Mont Blanc, Western Europe's highest point.

There was no nuance to this. The Alpine glaciers were dissolving. The snows that once seemed eternal – that had always been solid enough to support cable cars and mountain railways – were turning into mud.

And new evidence kept pouring in. As the mercury touched 40°C in that summer of 2022 half of Europe seemed to be on fire. It was the Year Without a Shower. Flames (some of them started deliberately) licked through dried-out woodland from Portugal to Hungary. And it was not only the mountains that were feeling the heat: the rivers were suffering too.

In August the water gauge at Kaub, near Mannheim (where Blücher chased Napoleon across the river) recorded a dangerously low depth of 16 inches, far below the 30 inches needed for container ships. Given the river's importance as a transport conduit for household names such as BASF and ThyssenKrupp, Deutsche Bank had to lower its predictions for German GDP. And that was not a first: the drought of 2018 had interrupted some €4 billion worth of river traffic, more than enough to put a dent (0.2 per cent) in the national economic performance. In that earlier drought it emerged that the Rhine was not just a vital artery for heavy industry: the majority of Germany's bread rolls were baked on these same riverbanks.

Was it ironic that the Rhine barges, which carried raw materials to the power plants that were actually contributing to this ecological crisis, were favoured by EU climate-change policies on the grounds that water-based freight was environmentally friendly? Perhaps. But the shortfall was so dramatic that a tragic object appeared in the shallows near Worms – one of the famous Hunger Stones placed in 1947 to commemorate the drought that afflicted this bomb-ruined patch of post-war Germany. It wasn't as old as the stone marker in the Elbe, which commemorated a nineteenth-century drought with the plaintive message: 'If you see me, weep.' But it was a blunt reminder of the fact that the low water of 2022 was no once-in-a-hundred-year anomaly. If anything, it was one of those one-offs that now seemed to be happening all the time.

There had been similar droughts in 2018, 2019 and 2020. These days, a year *without* a heatwave seems a newsworthy rarity. At Emmerich, near Arnhem (in that same summer of 2022) the Rhine was almost dry. Dog walkers in Bingen and Cologne could wander down the middle of their riverbed. The boats on Lake Constance lay tilted on mudflats. Nor was the Rhone immune. As France suffered what Prime Minister Élisabeth Borne was calling a 'special crisis' (all but three of France's ninety-six *départements* had to impose water restrictions), Électricité de France (EDF) had to cut output at its nuclear plants because the Rhone was too warm. Bugey, half an hour upriver from Lyon, has four nuclear reactors. Saint-Alban, south of Vienne and visible from the Côte-Rôtie, has two. There are more plants at Cruas, north of Montélimar, and Tricastin,* a few miles

---

* It used to be assumed that Tricastin meant 'three castles', like the town it faces, Saint-Paul-Trois-Châteaux. In fact it is the name that is the bowdlerisation: the Tricastini were a pre-Roman tribe, indigenous long before anyone thought of building castles.

south, which keep 6,000 people occupied running eight reactors delivering 10 per cent of France's total electricity needs.

All of these were imperilled. They are highly sensitive instruments, and the Rhone is crucial to their welfare: its fast-flowing water is their cooling system. It still generates hydroelectric power too. The barrage at Donzère, midway between Montélimar and Lyon, is a palatial bridge above a chute filled with water from the Rhone. The dam is 3 miles long, 180 feet high (a touch taller than Nelson's Column) and contains five turbines, using the magic of modern science to spin water into Wi-Fi.

So while the Rhone is a living product of French geology and history, it is a workhorse too. And as the glaciers wilt, that horse is becoming less dependable. In the heatwave of 2003, the temperature of the Rhone rose so sharply that the generators had to be paused, and in 2008 there was a leakage of radioactive uranium waste that led to a ban on drinking water, swimming and fishing. Wine sales dropped, and a number of the Côtes du Rhone vineyards that had the misfortune to feature the *pays d'origine* – Tricastin – on their labels quietly changed their name.

Ministers of energy all over Europe held their breath. They lived in a world of intricate co-dependencies, but the idea that Europe's rivers might run dry had not, until lately, been in their model. And they seemed to be facing an enemy that was attacking on two fronts. On the one hand, global warming was melting the glaciers, intensifying the rainstorms and causing worse and more frequent floods; on the other, it was creating heatwaves in which a long-established river could simply evaporate. In some ways the old wars, when the enemy was visible, had been easier.

The situation in Lombardy was just as severe. In that same drought of 2022, the canals out of the Ticino that usually irrigated the rice paddies were empty, and the authorities were reduced to trucking water down from Lake Maggiore, which itself was over 6 feet below its usual level (the lowest measurement since the 1940s). Zenne, an hour south of Milan, had its worst harvest since 1952, and, mindful of the 4,000 risotto farms in the area, the government declared a state of emergency.

The clams in the Po Delta were dying from lack of oxygen. *Spaghetti con vongole* was falling off the menus of Italian restaurants from Venice to Dublin. And the struts of sunken vessels were beginning to poke above the surface of the river like submarine periscopes – a relevant echo, since some of them were leftovers from the Second World War. Sappers even had to detonate an unexploded bomb that appeared on an exposed bank near Mantua. Reeds and grasses flourished on the muddy fringes, dividing a once enormous river into a fan of shallow streams.

The river's flow – sluggish at the best of times – became a crawl. In places the current was so weak that water from the Adriatic backed up towards Mantua, poisoning crops and reminding anyone with 200-year-old memories that in Napoleonic times this had been a poisonous swamp. Here again one could feel the past flowing into the present. The draining of the marshes only exposed new monsters.

The same is true all over Europe. The waters that stream out of the Alps – like wriggling lines of gunpowder, creating new worlds wherever they run – are increasingly vulnerable. Threatened from above by the changing climate, and from below by the contaminating impacts of modern life, during periods of drought they can become imperilled to an extent that puts our dependence on them in a new light. It will no longer suffice to see the Rhine, the Rhone and the Po as mere servants or beasts of burden. They are rich and mysterious organic systems with impulses and reflexes of their own. It is time to treat them as equals – perhaps even to protect them as if they were endangered species – not least because if we do refuse to change our ways, then the rivers might well change them for us.

34

BETTER TOGETHER

In 2004 the European Union published a map that included its newest members: Cyprus, the Czech Republic, Hungary, Malta, Poland, Slovakia, Slovenia and the Baltic States. These accessions had enlarged the EU's population to 450 million.

There was something curious about this expansion, though. Something that had originally been conceived as a coal and steel pact between France and Germany, and had then evolved into a free-trade area with common laws, now bore a remarkable resemblance to the great Carolingian Empire of the ninth century. Like Charlemagne's domain, it stretched from the Pyrenees to Russia, and from the Baltic to Rome.

'Charlemagne would recognise the geography,' wrote Vincent Virga in *Cartographia*, a history of maps and mapmakers published in 2007. The capitals of modern Europe – Brussels, Frankfurt, Strasbourg – had also lain at 'the heart of his realm' in the Rhineland. His sweeping imperium had extended down the Rhone and Po as well. It could have been called the Empire of the Three Rivers.

Hindsight is easy. Only now can we see that Charlemagne was planting and spreading an idea – a united continent in Western Europe – that was at least a thousand years ahead of its time.

It did not take long for that tentative union to break apart: when Charlemagne died, his blueprint for a joined-up continent splintered into the three tranches inherited by his sons – west of the Rhone (France), the terrain between Rhone and Rhine (Burgundy), and east of the Rhine (Germany). The central section, Franco-German-Italy, became first Lotharingia and then Lorraine, and was thus the cause of near-continuous war for the next millennium. Not until the terrible wars of the twentieth century was Europe shocked into attempting something new. Finally, it sought a different path. Charlemagne's empire had been created by war; now, with nuclear oblivion a distinct possibility, the modern union was an attempt to create unity by peaceful means.

It has been said that history does not really recur – it plays with resemblances. Yet in purely topographical terms, as that 2004 map made clear, the modern dream of European togetherness was in one sense not a surprising departure but a giant leap backwards. The continent of the three rivers had been trying (and failing) to regain that lost unity for a thousand years. So while in practice modern Europe may be very different from anything Charlemagne would have conceived, the overall vision is in some ways similar.

Europe's past is a tangled one, a blurred jumble of contradictory images. From one angle it is a doleful pageant of swords and cannons, bombs and concentration camps; from another, it is a blizzard of breakthroughs and masterpieces. One can never overlook the nightmares in its past; but these have almost always been spliced with miracles.

Another way of putting this dichotomy is to see Europe as having always had two equal and opposite qualities: variety and homogeneity. In some lights its differences are clearly demarcated – in language, music, food, and all other aspects of its cultures. But it is also brought together by the extent of what it shares. This is no longer anything to do with its being 'Christendom': religious uniformity splintered centuries ago, and in many places is a dwindling force. But there remains a recognisable consistency in the way its different pieces eat, legislate, drink, pray, think and love.

As Stefan Zweig wrote: 'If a forgotten comedy by Terence is found in a hidden corner of Italy, there is a cry of joy in England, as in Poland and Spain . . . as if a child had been born, or fortune dropped from the sky.' In this light history seems most of all like an attempt to defy the dictates of geography. The holy grail of peace and unity may have been mostly a mirage, but Europe has never called off the search.

One of the hopes of history is that peering into the past might give us a better view of the future. That is not easy: the future is a closed book, and the past can be just as cloudy. Up where our rivers began, I imagined that I had glimpsed the power of the water cycle, and from that vantage point the rivers seemed centrifugal, binding Europe together. Now, in the lowlands, they looked very different: a centripetal arrangement that sought to tug Europe's realms apart – to emphasise difference.

But it is in the nature of rivers to contradict themselves. They erode *and* deposit, create *and* destroy, divide *and* unite. They have been the cause of many wars but, no matter how sharp the differences in the lands through which they run, they have also inspired ideas of compromise and co-operation. It was the need for flood protection that knocked the Swiss cantons into a confederacy. Across the continent, villages collaborated over water for their mills and fishing rights. Cities set aside their rivalries to do the same. Tutored by floods and droughts, nations created new offices to supervise the rivers they shared. Even Napoleon, not averse to settling disputes with muskets, saw that it was in France's interest that the Rhine be a brotherhood. For centuries, these rivers have taught Europe both the price of conflict and the bottomless value of peace.

In 1870, after the Franco-Prussian war, Victor Hugo predicted (and pleaded) that when his country recovered it would do so not in anger but in friendship: 'No more frontiers! The Rhine for everyone! Let us be the same Republic, let us be the United States of Europe, let us be the continental federation, let us be European liberty, let us be universal peace.'

Winston Churchill is usually credited with having been the first to voice this concept of a 'United States of Europe', in a 1946 speech in Zurich. But the author of *Les Misérables* had used the exact same phrase almost a century earlier. And it was not his first such declaration. Back in 1849 he had called for a more unified Europe by telling the Paris Peace Congress: 'A day will come when the only fields of battle will be markets opening up to trade, and minds opening up to ideas.'*

Those words today adorn the bedroom of his house on the Place des Vosges in Paris, now a museum. And perhaps they summarise the lesson of this ramble down Europe's waterways. By marrying the landscapes through which they pass to the events they have witnessed, the rivers have stitched the continent's geography and history into a single pattern. Racing away from each other, they remain part

---

* The Paris Congress, chaired by Hugo, was part of a series of such meetings. Others were held in London, Brussels and Frankfurt, with speeches by the English author and MP Richard Cobden and the American abolitionist and former slave William Wells Brown.

of a great system, connected at the top, that has nourished Europe all its life.

~

There are three endings to this journey.

In the south of France, the Rhone starts to feel like the sea – not yet salt, but sluggish as it eases into the mudflats of the Camargue – as soon as it leaves Arles. The fact that this is where the river forks made it a natural stronghold in classical times: it has one of the finest sets of Roman remains in all of Europe, and was the birthplace of the Emperor Constantine's son. Later it was the bread capital of Provence: the mill complex a mile outside the town had sixteen wheels that produced five tonnes of flour per day. It was a seaport, too, and though the Rhone has filled the shallows with silt, it still feels like a fortified harbour.

Back then it protected one end of the first Roman road in this region, the Via Aurelia, which ran through southern France and on to Sicily. Perhaps people thought, as the street drove through Salon-de-Provence (once home to Nostradamus, now to an air force base – which he did not predict) and then Antibes, Nice, Genoa, Piacenza and all stations to Rome, that all roads led to Arles.

From here the river divides into two. The Grand Rhone is the shipping lane to Port-Saint-Louis (dug in 1871), and not a place to linger unless you love salt pans, oil terminals, petrochemical factories and grain silos. The western branch, the Petit Rhone, is different, an unspoiled wetland whose flamingos and wild horses give it the appearance of another continent. On its margin stands the walled city of Aigues-Mortes, a medieval garrison on land that was not even land in Roman times, looking out over the salt flats of the Camargue like a marooned castle. Charlemagne built a watch tower here in 791 CE, and it was an assembly point during the crusades: Louis IX set sail from here twice. The name of this delta is the 'Rhone Fan', and it is part of the municipal commune known as Bouches-du-Rhone. This is a river-made world.

It is a very different story for the Rhine. This peters out in the Netherlands, fanning into the Rhine–Meuse–Scheldt delta. The Rijn itself is a minor channel to the Hook of Holland; it is the Waal and the Scheldt that become the main (and enormous) shipping routes to Dordrecht and Rotterdam. These are the waters that made Holland wealthy in the seventeenth century, and give it its substance today. There are artistic echoes (in the name of Rembrandt van Rijn, for instance, whose very name reminds us that the entire Dutch Golden Age was given much of its gilding by this ocean-going superhighway) but the lower Rhine is above all a marine heavyweight. Rotterdam is Europe's biggest port, and not far off being the largest in the world (it was displaced by Singapore in 2004). It handles 30,000 ocean-going ships a year, and some 100,000 river vessels, which between them shift 500 million tonnes of freight per annum. Container ports, oil depots, docks, petrochemical wharves – the Europoort is no one's idea of a tourist gem, but it is one of Europe's most important interchanges – and most dramatic sights.

The Po ends much more quietly. It flops into a marsh and UNESCO-protected nature reserve, dazed by a sweltering sun. Wary wading birds paddle in the rich shallows, looking for eels, clams and shrimp, while the brick faces of Etruscan farmhouses squat in the shade of pine trees or gleam across the reed banks. The Po has fourteen mouths, but only two of them are significant routes into the Adriatic. The 'active' delta is a relatively recent invention, dug by the Venetian Republic to keep silt away from its approaches.

All roads lead to the sea: Europe's rivers are a precious
inheritance we must not take for granted.

The original or 'fossil' delta, on the other hand, is a brackish dead end. This is where the pollutants from Turin and Milan end up – fertilisers, plastics, drugs, sewage, rubbish waste and oil spills.

All three of these endings are a long way from an Alpine peak. But the deltas look in two directions at once – out towards the ocean lanes and back to the mountains behind. In the tension between this divergence crackles the mixture of creativity and misery that has been Europe's enduring watermark. It flows from the simple geographical fact that its rivers are always more than one thing. Borders? Arteries? Symbols? Taps? Highways? Drains? Weak points? Hazards? They are all of these – and also an infinite work in progress. In 2024 work began on a multi-million-euro project to make the artificially rocky banks of the Rhone south of Lyon more porous, to regenerate the wetland ecosystem that had dried out and withered over the years. Similar schemes are ongoing along both the Rhine and the Po; it seems there is no end to our attempts to correct what nature wrote.

All three waters were born in the same nursery, raised in the same cradle, united by memories of a common infancy, hurtling their way to the sea like heedless children, arguing, drifting, racing, twisting, surging – yet always there. Change and continuity, locked in an endless dance.

High above Andermatt, meanwhile, the Rhine, the Rhone and the Ticino quietly drip out of their glacial nests. They may not wander where they please: their destinies have already been arranged. But the seas into which they pour will soon send those waters into the sky, to drift back over the mountains and begin their roundabout journey all over again.

ACKNOWLEDGEMENTS AND SELECT BIBLIOGRAPHY

The following is a list of the books I found most useful. But thanks are also due to innumerable libraries, museums, newspapers and tourist offices all over France, Germany, Italy and Switzerland. And then there is the smartphone, which has put an encyclopaedia into the backpack of every rambler – a thousand thanks to all the online resources. One that is not thanked as often as it should be, perhaps, is the most obvious – Wikipedia – for where else, when you are halfway up a mountain or walking along a river, can you check the exact length of the Rhine or the price of coal in the nineteenth century, track down a Rilke poem, look up the history of an unusual Rhone vintage, glance at a Van Gogh painting, pin down the location of a battlefield, rustle up a note on a Roman wall or Gothic spire, find the spot where some little-known engineer was buried, bone up on pollution levels, flood zones or glacial collapses, and explore a hundred other urgent matters – all in the time it takes to order a cup of coffee?

Most of all I must mention the kind and expert people who helped bring the book to fruition. Pippa Crane, Celia Hayley and Jill Burrows spared me a thousand blushes, and did not complain that I imposed on them so heavily. David Godwin was the perfect steward and sounding board ('sounds great . . . just keep going.'). And Hermione Davies was, as always, there every step of the way, sometimes far ahead. Heartfelt thanks to all.

Acemoglu, Damon and Robinson, James A., *Why Nations Fail* (Profile, 2012)

Barzun, Jacques, *From Dawn to Decadence: 500 Years of Western Cultural Life* (HarperCollins, 2000)

Beattie, Andrew, *The Alps: A Cultural History* (Signal, 2006)

Bewes, Diccon, *Swiss Watching* (Nicholas Brealey, 2018)

Black, Jeremy, *The British Abroad: The Grand Tour in the Eighteenth Century* (Sutton, 1992)

Blackbourn, David, *The Conquest of Nature: Water, Landscape and the Making of Modern Germany* (Cape, 2006)

Braudel, Fernand, *A History of Civilisations* (Penguin, 1993)

Cioc, Mark, *The Rhine: An Eco-biography* (University of Washington, 2006)

Clark, Ronald W., *The Alps* (Weidenfeld & Nicolson, 1973)

Coates, Ben, *The Rhine* (Nicholas Brealey, 2019)

Cole, Robert, *A Traveller's History of France* (Windrush, 1988)

Coolidge, W. A. B., *The Alps in Nature and History* (Methuen, 1908)

Dumas, Alexandre, *Travels in Switzerland* (Peter Owen, 1958)

Durrell, Lawrence, *Spirit of Place* (Faber & Faber, 1988)

Egli, Emil, *Swiss Life and Landscape* (Paul Elek, 1949)

Fahrni, Dieter, *An Outline History of Switzerland* (Arts Council of Switzerland, 2003)

Fenby, Jonathan, *The History of Modern France* (Simon & Schuster, 2015)

Gardner, Roy, Gaston, Noel and Masson, Robert T., *Tolling the Rhine in 1254* (Indiana University, 2002)

Gould, Stephen Jay, *Time's Arrow, Time's Cycle* (Pelican, 1988)

Greengrass, Mark, *Christendom Destroyed: Europe 1517–1648* (Penguin, 2015)

Hawes, James, *The Shortest History of Germany* (Old Street, 2017)

Hinsdale, Jeremy, *The Culture and History of Glaciers in the Alps* (Columbia, 2018)

Hugo, Victor, *The Rhine* (London, 1843)

Jones, Tobias, *The Po: An Elegy* (Head of Zeus, 2022)

Landes, David, *The Wealth and Poverty of Nations* (Abacus, 1999)

Leonardo da Vinci, *The Notebooks* (Oxford, 1952)

Levin, Bernard, *To the End of the Rhine* (Cape, 1989)

Ludwig, Emil, *The Nile: The Life Story of a River* (Allen and Unwin, 1936)

Maddocks, Fiona, *Hildegard of Bingen: The Woman of Her Age* (Faber & Faber, 2001)

Man, John, *The Gutenberg Revolution* (Headline, 2002)

Magris, Claudio, *Danube* (Harvill, 1986)

Marshall, Tim, *The Power of Geography* (Elliott & Thompson, 2021)

Mauch, Christof, and Zeller, Thomas G. (eds), *Rivers in History* (University of Pittsburgh Press, 2008)

Morris, Jan, *Fifty Years of Europe: An Album* (Viking, 1997)

Morris, Jan, *A Writer's World* (Faber & Faber, 2003)

Nicholl, Charles, *Leonardo da Vinci* (Penguin, 2004)

O'Shea, Stephen, *The Alps: A Human History* (Norton, 2017)

Pettegree, Andrew, *The Invention of News* (Yale, 2014)

Rady, Martin, *The Habsburgs* (Penguin, 2020)

Robb, Graham, *The Discovery of France* (Picador, 2007)

Rooney, Padraig, *The Gilded Chalet* (Nicholas Brealey, 2017)

Russell, John, *Switzerland* (Batsford, 1950)

Saltzman, Cynthia, *Napoleon's Plunder, and the Theft of Veronese's Feast* (Farrar, Straus and Giroux, 2021)

Schlee, Ann, *Rhine Journey* (Pan Macmillan, 1981)

Spitteler, Carl, *Der Gotthard* (Gotthard Railway Company, 1897)

Steinberg, Jonathan, *Why Switzerland* (Cambridge, 1996)

Stephen, Leslie, *The Playground of Europe* (Longman, 1894)

Taylor, Andrew, *The World of Gerard Mercator* (HarperCollins, 2005)

The Times Atlas of European History (HarperCollins, 1994)

Tuchman, Barbara, *The March of Folly* (Abacus, 1985)

Virga, Vincent, *Cartographia: Mapping Civilisations* (Little Brown, 2007)

Wedgwood, C. V., *The Thirty Years War* (New York Review, 2005)

Winder, Simon, *Germania* (Picador, 2010)

Woodward, Jamie, *The Ice Age* (OUP, 2014)

INDEX

INDEX